Role of TCTP in Cell Biological and Disease Processes

Role of TCTP in Cell Biological and Disease Processes

Editors

Ulrich-Axel Bommer
Toshiaki Kawakami

MDPI • Basel • Beijing • Wuhan • Barcelona • Belgrade • Manchester • Tokyo • Cluj • Tianjin

Editors
Ulrich-Axel Bommer
University of Wollongong
Australia

Toshiaki Kawakami
La Jolla Institute for Immunology
USA

Editorial Office
MDPI
St. Alban-Anlage 66
4052 Basel, Switzerland

This is a reprint of articles from the Special Issue published online in the open access journal *Cells* (ISSN 2073-4409) (available at: https://www.mdpi.com/journal/cells/special_issues/TCTP).

For citation purposes, cite each article independently as indicated on the article page online and as indicated below:

LastName, A.A.; LastName, B.B.; LastName, C.C. Article Title. *Journal Name* **Year**, *Volume Number*, Page Range.

ISBN 978-3-0365-2285-2 (Hbk)
ISBN 978-3-0365-2286-9 (PDF)

Contents

About the Editors

Ulrich-Axel Bommer, Clinical Associate Professor at the School of Medicine, University of Wollongong, NSW, Australia. He received his Ph.D. degree from the Martin-Luther-University, Halle, Germany. As postdoctoral fellow at the Institute of Molecular Biology, East German Academy of Sciences, Berlin-Buch, he studied the structure and function of ribosomal complexes, which provided the basis for his Habilitation at the Martin-Luther-University, Halle. From 1991, he worked as a lecturer/senior lecturer at St. George's Hospital Medical School in London (UK), where he began research on TCTP. From 2006 to 2014, he was Associate Professor and a founding member at the Graduate School of Medicine, University of Wollongong, Australia.

Toshiaki Kawakami, Professor at the La Jolla Institute for Immunology, La Jolla, California, USA, received his M.D. and Ph.D. degrees from The University of Tokyo, Faculty of Medicine, Tokyo, Japan. He studied oncogenes as a postdoctoral fellow at the National Cancer Institute/NIH, Bethesda, Maryland, USA. After being Assistant Professor at Kyoto University, he joined the newly formed La Jolla Institute for Allergy and Immunology (currently La Jolla Institute for Immunology) as an assistant member (equivalent to Assistant Professor) in 1990 to start mast cell and allergy research.

cells MDPI

Editorial

Role of TCTP in Cell Biological and Disease Processes

Ulrich-Axel Bommer [1,*] and Toshiaki Kawakami [2,*]

1 School of Medicine, Faculty of Science, Medicine and Health, University of Wollongong, Northfields Avenue, Wollongong, NSW 2522, Australia
2 Laboratory of Allergic Diseases, Center for Autoimmunity and Inflammation, La Jolla Institute for Immunology, 9420 Athena Circle, La Jolla, CA 92037, USA
* Correspondence: ubommer@uow.edu.au (U.-A.B.); toshi@lji.org (T.K.)

Translationally controlled tumor protein (TCTP), also referred to as histamine-releasing factor (HRF) or fortilin, is a multifunctional protein, expressed in essentially all eukaryotic organisms. TCTP is involved in many basic biological processes, such as stress responses, cell division, as well as growth and development, both at the cellular and the organ level. It is therefore not surprising that the dysregulation of TCTP occurs in various disease processes, such as cardiovascular, allergic and immune disorders. TCTP's role in cancer-promoting pathways is particularly well documented, and the protein is considered to be a potential target for the design of new anti-cancer strategies. Therefore, an understanding of the core biological functions of TCTP, the mechanisms underlying its cellular regulation, and its involvement in disease processes, is essential. With this goal in mind, together with the contribution of all the authors, we compiled the Special Issue 'Role of TCTP in Cell Biological and Disease Processes' in *Cells*, the articles of which comprise this book. Through the inclusion of three review articles, we aimed to provide a current overview on a wide range of aspects pertaining to this protein, and four original papers highlight some recent developments in this area.

Citation: Bommer, U.-A.; Kawakami, T. Role of TCTP in Cell Biological and Disease Processes. *Cells* **2021**, *10*, 2290. https://doi.org/10.3390/cells10092290

Received: 23 August 2021
Accepted: 27 August 2021
Published: 2 September 2021

Involvement of TCTP in Core Biological Functions

Three of these original papers reported examples of core biological processes, in which TCTP is involved. The study by S-H Chen and co-workers [1] investigated the importance of TCTP for the development of the central nervous system in mice. This group generated a conditional knockout mouse, where TCTP was disrupted in neuronal and glial progenitor cells. These mice died at the perinatal stage, and the results indicated that TCTP is a critical protein for cell survival during early neuronal differentiation.

Maria Lucibello and colleagues studied the importance of TCTP for cell division in human breast cancer cells [2]. It was known for quite some time that TCTP can be found to be associated with the mitotic spindle [3], and that it undergoes mitotic phosphorylation by the polo-like kinase Plk-1 [4], but the involvement of this process in breast cancer was only investigated much later. In their previous paper, Lucibello and co-workers showed that phospho-TCTP levels are particularly elevated in aggressive breast cancer [5]. Here, they demonstrate that phospho-TCTP is essential for correct transition through mitosis in human mammary epithelial cells.

Autophagy, a lysosomal degradation pathway, is another core biological process, which is important for maintaining homeostasis at the cellular, tissue and organism level. Only a few papers on TCTP and autophagy have been published so far, and the regulatory role of the protein in autophagy is still controversial, since both stimulation and inhibition of autophagy by TCTP were observed. Therefore, we welcomed the review article by Kyunglim Lee and colleagues [6], which discussed these aspects of TCTP function, in relation to its role in tumorigenesis. An original study by Vojtova and Hasek [7] investigated the importance of Mmi1, the yeast orthologue of mammalian TCTP, in non-selective autophagy in the yeast *S. cerevisiae*. They found that Mmi1 negatively affects rapamycin-induced autophagy, whereas it had no effect on nitrogen starvation-induced autophagy.

Our own review article [8] aimed to provide a comprehensive overview on the core biological functions of TCTP. Apart from the above-mentioned processes of cell/organ development and autophagy, it listed regulation of protein synthesis, stability regulation of key proteins, and biological stress responses as the main groups of cell biological processes, in which TCTP may be involved. Various interesting examples of TCTP's activities for each of these categories can be found in this article. A general overview on the biological activities of TCTP is schematically summarized in Figure 1 (green and orange boxes). The figure is largely based on the data compiled in the review articles of this Special Issue and, apart from listing the core biological activities of TCTP, it also summarizes the main groups of disease processes, where TCTP/HRF/fortilin or its dysregulation was reported to play an important role (grey boxes). Our review article cites all the other articles of this Special Issue, where relevant, thereby putting them into context. In addition, our review provides an account of the various regulatory mechanisms that are involved in regulating cellular TCTP levels, and which are important for the understanding of some aspects of TCTP dysregulation in diseases.

Disease Processes in Which TCTP Participates

The role of TCTP in cancer is well documented in previous reviews (for a collection of such articles, see [9]); therefore, in our review [8], we focused on the participation of TCTP in the most important cancer cell biological processes, as they are listed in the grey box in Figure 1. The above-mentioned article by Lucibello and colleagues [2] represents an original contribution to the topic 'TCTP and cancer'. The authors addressed the question as to whether phospho-TCTP might be a suitable target for anti-cancer treatment strategies in aggressive breast cancer. They showed that treatment with dihydro-artemisinin (DHA) resulted in a reduction in phospho-TCTP levels and caused mitotic aberration in trastuzumab-resistant breast cancer cells. They also demonstrated, in an orthotopic breast cancer xenograft model that, in combinatorial treatment, DHA improved the long-term efficacy of trastuzumab emtansine (T-DM1) considerably, suggesting that targeting phospho-TCTP might be a suitable element for a combinatorial anti-cancer strategy in this type of breast cancer. The review article by Lee and colleagues [6] summarized various additional aspects of TCTP's participation in cancer, with an emphasis on anti-apoptosis, cancer cell invasion, resistance to anti-cancer therapies, as well as the potential of TCTP as a prognostic tool in cancer.

Another group of disorders, where TCTP dysregulation was shown to be involved, are cardiovascular and metabolic diseases, such as atherosclerosis, hypertension, and diabetes (grey box in Figure 1). These observations are reviewed in our article [8] (for an earlier review also see [10], where TCTP is referred to as fortilin). An important general aspect here is that, on one hand, TCTP is important for the survival of specialized cell types, such as pancreatic β cells [11] and cardiomyocytes [12]; on the other hand, excess TCTP may lead to hypertrophy, as shown for nephrotic podocytes in diabetes [13] or in skeletal muscle [14].

Since the discovery of the histamine-releasing factor (HRF) and its molecular identification as an extracellular form of TCTP [15], it is known that TCTP/HRF is also involved in allergic and immune disorders. Unravelling the precise role of the (dimeric) extracellular form of HRF in these processes took considerable research effort, spread over more than 20 years [16]. The interesting recent developments in this area are summarized in the review article by Kawakami et al. [17] (also see the grey box in Figure 1). Another example of an extracellular activity of TCTP/HRF was revealed by the discovery that the protein is a component present in the venom of the brown spider *Loxosceles intermedia* [18]. In an effort to characterize the importance of TCTP/HRF in the spider's venom, Senff-Ribeiro and co-workers studied the pro-inflammatory activity of TCTP in Loxoscelism, which is the reaction that develops after a bite by this spider [19]. They concluded that TCTP is an essential synergistic factor for the dermo-necrotic actions of the main toxin contained in the venom.

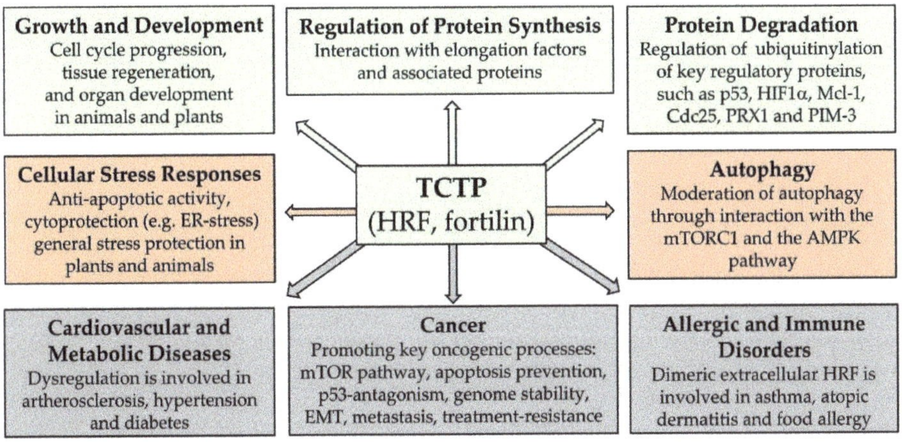

Figure 1. Schematic summary of core biological processes and of principal disease processes where TCTP is involved. (Green boxes: cell regulatory processes promoting growth and proliferation; orange boxes: cellular maintenance and defense systems; grey boxes: disease processes). This summary scheme is based on data compiled in the review articles [6,8,17], but also incorporates findings reported in some of the original articles of this Special Issue.

The extracellular function of TCTP as a histamine releasing factor, in the context of allergic reactions, was originally discovered by Susan MacDonald [15]. She and her team made a considerable contribution to elucidating several aspects related to the activity of HRF. Sadly, we learned that in September 2020, Dr. MacDonald passed away. Given her original discovery of the HRF and her further contribution to the field, we considered it appropriate to include an obituary for her in this Special Issue on TCTP/HRF [20].

Recent Developments

Since the completion of our Special Issue, a few interesting biological activities of TCTP have been reported in non-vertebrate organisms. Some examples are its role in the growth, development and differentiation of the slime mold *Dictyostelium discoideum* [21]; the importance in *Drosophila* for epithelial integrity and organ growth [22]; and in *Toxoplasma gondii* for robust growth of the parasite and for the maintenance of its virulence [23]. Recognizing the importance of TCTP for the survival of unicellular parasites, Bossard et al. [24] took the step to test the efficacy of immunizing mice against the parasitic TCTP of *Trypanosoma brucei*. The immunized mice displayed a reduced first peak of parasitemia, a two-fold delay in the onset of the second peak and an increased time of survival, compared to the control animals.

The group of Kyunglim Lee investigated the metabolic importance of TCTP in mice. They generated a TCTP-overexpressing mouse and found that these display an improved metabolic homeostasis, with enhanced glucose tolerance, insulin sensitivity, and energy expenditure, compared to control mice [25]. In another paper, the group studied the importance of TCTP in rheumatoid arthritis (RA) [26]. They found that TCTP levels are increased in the sera and synovial fluids of patients with RA, compared to control groups. Their results indicate that TCTP might serve as a biomarker and a therapeutic target in RA patients. A novel tumor–promoting role of TCTP has just been published in Nature Immunology [27]. The authors observed that the release of TCTP from necrotic cancer cells switches on an immunosuppressive network of myeloid-derived suppressor cells, which substantially contributes to the suppression of antitumor immunity, and thus to tumor progression. All these examples show that we can still expect new facets of TCTP's role to be unraveled, both regarding its biological and its disease-promoting activities.

Funding: U.-A.B. received small grants from the Illawarra Health and Medical Research Institut University of Wollongong. T.K. was supported by the US National Institutes of Health gran AI146042 and AI153867.

Acknowledgments: We thank all authors who contributed to this Special Issue. The guidance b Janet Yan and her colleagues through the editorial process is highly appreciated. We thank th Editorial team of *Cells* for the opportunity to serve as Guest Editors for this Special Issue.

Conflicts of Interest: The authors declare no conflict of interest.

References

1. Chen, S.H.; Lu, C.H.; Tsai, M.J. TCTP is Essential for Cell Proliferation and Survival during CNS Development. *Cells* **2020**, *9*, 13 [CrossRef] [PubMed]
2. D'Amico, S.; Krasnowska, E.K.; Manni, I.; Toietta, G.; Baldari, S.; Piaggio, G.; Ranalli, M.; Gambacurta, A.; Vernieri, C.; Di Giacint F.; et al. DHA Affects Microtubule Dynamics Through Reduction of Phospho-TCTP Levels and Enhances the Antiproliferativ Effect of T-DM1 in Trastuzumab-Resistant HER2-Positive Breast Cancer Cell Lines. *Cells* **2020**, *9*, 1260. [CrossRef]
3. Gachet, Y.; Tournier, S.; Lee, M.; Lazaris-Karatzas, A.; Poulton, T.; Bommer, U.A. The growth-related, translationally controlle protein P23 has properties of a tubulin binding protein and associates transiently with microtubules during the cell cycle. *J. Ce Sci.* **1999**, *112*, 1257–1271. [CrossRef] [PubMed]
4. Yarm, F.R. Plk phosphorylation regulates the microtubule-stabilizing protein TCTP. *Mol. Cell Biol.* **2002**, *22*, 6209–6221. [CrossRe
5. Lucibello, M.; Adanti, S.; Antelmi, E.; Dezi, D.; Ciafre, S.; Carcangiu, M.L.; Zonfrillo, M.; Nicotera, G.; Sica, L.; De Brau F.; et al. Phospho-TCTP as a therapeutic target of Dihydroartemisinin for aggressive breast cancer cells. *Oncotarget* **2015**, 5275–5291. [CrossRef]
6. Lee, J.S.; Jang, E.H.; Woo, H.A.; Lee, K. Regulation of Autophagy Is a Novel Tumorigenesis-Related Activity of Multifunctiona Translationally Controlled Tumor Protein. *Cells* **2020**, *9*, 257. [CrossRef] [PubMed]
7. Vojtova, J.; Hasek, J. Mmi1, the Yeast Ortholog of Mammalian Translationally Controlled Tumor Protein (TCTP), Negativel Affects Rapamycin-Induced Autophagy in Post-Diauxic Growth Phase. *Cells* **2020**, *9*, 138. [CrossRef] [PubMed]
8. Bommer, U.A.; Telerman, A. Dysregulation of TCTP in Biological Processes and Diseases. *Cells* **2020**, *9*, 1632. [CrossRef]
9. Telerman, A.; Amson, R. (Eds.) *TCTP/tpt1-Remodeling Signaling from Stem Cell to Disease*; Springer: Berlin/Heidelberg, German 2017; Volume 64, p. 309.
10. Pinkaew, D.; Fujise, K. Fortilin: A Potential Target for the Prevention and Treatment of Human Diseases. *Adv. Clin. Chem.* **201** *82*, 265–300. [CrossRef] [PubMed]
11. Tsai, M.J.; Yang-Yen, H.F.; Chiang, M.K.; Wang, M.J.; Wu, S.S.; Chen, S.H. TCTP is essential for beta-cell proliferation an mass expansion during development and beta-cell adaptation in response to insulin resistance. *Endocrinology* **2014**, *15* 392–404. [CrossRef]
12. Cai, W.; Fujita, T.; Hidaka, Y.; Jin, H.; Suita, K.; Shigeta, M.; Kiyonari, H.; Umemura, M.; Yokoyama, U.; Sadoshima, J.; et a Translationally controlled tumor protein (TCTP) plays a pivotal role in cardiomyocyte survival through a Bnip3-dependen mechanism. *Cell Death Dis.* **2019**, *10*, 1–14. [CrossRef]
13. Kim, D.K.; Nam, B.Y.; Li, J.J.; Park, J.T.; Lee, S.H.; Kim, D.H.; Kim, J.Y.; Kang, H.Y.; Han, S.H.; Yoo, T.H.; et al. Translationall controlled tumour protein is associated with podocyte hypertrophy in a mouse model of type 1 diabetes. *Diabetologia* **2012**, *5* 1205–1217. [CrossRef] [PubMed]
14. Goodman, C.A.; Coenen, A.M.; Frey, J.W.; You, J.S.; Barker, R.G.; Frankish, B.P.; Murphy, R.M.; Hornberger, T.A. Insights into th role and regulation of TCTP in skeletal muscle. *Oncotarget* **2017**, *8*, 18754–18772. [CrossRef] [PubMed]
15. MacDonald, S.M.; Rafnar, T.; Langdon, J.; Lichtenstein, L.M. Molecular identification of an IgE-dependent histamine-releasin factor. *Science* **1995**, *269*, 688–690. [CrossRef] [PubMed]
16. MacDonald, S.M. History of Histamine-Releasing Factor (HRF)/Translationally Controlled Tumor Protein (TCTP) Including Potential Therapeutic Target in Asthma and Allergy. *Results Probl. Cell Differ.* **2017**, *64*, 291–308. [CrossRef]
17. Kawakami, Y.; Kasakura, K.; Kawakami, T. Histamine-Releasing Factor, a New Therapeutic Target in Allergic Diseases. *Cells* **201** *8*, 1515. [CrossRef] [PubMed]
18. Sade, Y.B.; Boia-Ferreira, M.; Gremski, L.H.; da Silveira, R.B.; Gremski, W.; Senff-Ribeiro, A.; Chaim, O.M.; Veiga, S.S. Molecula cloning, heterologous expression and functional characterization of a novel translationally-controlled tumor protein (TCTP family member from Loxosceles intermedia (brown spider) venom. *Int. J. Biochem. Cell Biol.* **2012**, *44*, 170–177. [CrossRef]
19. Boia-Ferreira, M.; Moreno, K.G.; Basilio, A.B.C.; da Silva, L.P.; Vuitika, L.; Soley, B.; Wille, A.C.M.; Donatti, L.; Barbaro, K.C Chaim, O.M.; et al. TCTP from Loxosceles Intermedia (Brown Spider) Venom Contributes to the Allergic and Inflammator Response of Cutaneous Loxoscelism. *Cells* **2019**, *8*, 1489. [CrossRef]
20. Togias, A.; Plaut, M.; Langdon, J.; Bommer, U.A.; Telerman, A. Obituary for Susan, M. MacDonald, M.D. *Cells* **2021**, *1C* 87. [CrossRef]
21. Kumar, R.; Maurya, R.; Saran, S. Investigating the Role of Translationally Control Tumor Protein in Growth, Development an Differentiation of Dictyostelium discoideum. *Front. Cell Dev. Biol.* **2020**, *8*, 742. [CrossRef] [PubMed]

22. Lee, S.R.; Hong, S.T.; Choi, K.W. Regulation of epithelial integrity and organ growth by Tctp and Coracle in Drosophila. *PLoS Genet.* **2020**, *16*, e1008885. [CrossRef]
23. Zheng, J.; Chen, Y.; Li, Z.; Cao, S.; Zhang, Z.; Jia, H. Translationally controlled tumor protein is required for the fast growth of Toxoplasma gondii and maintenance of its intracellular development. *FASEB J.* **2018**, *32*, 906–919. [CrossRef] [PubMed]
24. Bossard, G.; Rodrigues, V.; Tour, E.; Geiger, A. Mice immunization with Trypanosoma brucei gambiense translationally controlled tumor protein modulates immunoglobulin and cytokine production, as well as parasitaemia and mice survival after challenge with the parasite. *Infect. Genet. Evol.* **2021**, *87*, 104636. [CrossRef]
25. Jeon, Y.; Choi, J.Y.; Jang, E.H.; Seong, J.K.; Lee, K. Overexpression of translationally controlled tumor protein ameliorates metabolic imbalance and increases energy expenditure in mice. *Int. J. Obes.* **2021**, *45*, 1576–1587. [CrossRef] [PubMed]
26. Kim, M.; Choe, Y.; Lee, H.; Jeon, M.G.; Park, J.H.; Noh, H.S.; Cheon, Y.H.; Park, H.J.; Park, J.; Shin, S.J.; et al. Blockade of translationally controlled tumor protein attenuated the aggressiveness of fibroblast-like synoviocytes and ameliorated collagen-induced arthritis. *Exp. Mol. Med.* **2021**, *53*, 67–80. [CrossRef] [PubMed]
27. Hangai, S.; Kawamura, T.; Kimura, Y.; Chang, C.Y.; Hibino, S.; Yamamoto, D.; Nakai, Y.; Tateishi, R.; Oshima, M.; Oshima, H.; et al. Orchestration of myeloid-derived suppressor cells in the tumor microenvironment by ubiquitous cellular protein TCTP released by tumor cells. *Nat. Immunol.* **2021**, *22*, 947–957. [CrossRef]

cells

MDPI

Article

TCTP Is Essential for Cell Proliferation and Survival during CNS Development

Sung-Ho Chen [1,2,*], Chin-Hung Lu [2] and Ming-Jen Tsai [3]

1 Department of Pharmacology, College of Medicine, Tzu Chi University, Hualien 97004, Taiwan
2 Master Program in Pharmacology and Toxicology, College of Medicine, Tzu Chi University, Hualien 97004, Taiwan; Lu721112@gmail.com
3 Department of Emergency Medicine, Ditmanson Medical Foundation Chia-Yi Christian Hospital, Chiayi City 600, Taiwan; tshi33@gmail.com
* Correspondence: chensho@mail.tcu.edu.tw; Tel.: +886-3-8565301 (ext. 2452); Fax: +886-3-8561465

Received: 30 September 2019; Accepted: 2 January 2020; Published: 6 January 2020

Abstract: Translationally controlled tumor-associated protein (TCTP) has been implicated in cell growth, proliferation, and apoptosis through interacting proteins. Although TCTP is expressed abundantly in the mouse brain, little is known regarding its role in the neurogenesis of the nervous system. We used *Nestin-cre*-driven gene-mutated mice to investigate the function of TCTP in the nervous system. The mice carrying disrupted TCTP in neuronal and glial progenitor cells died at the perinatal stage. The $Nestin^{Cre/+}$; $TCTP^{f/f}$ pups displayed reduced body size at postnatal day 0.5 (P0.5) and a lack of milk in the stomach compared with littermate controls. In addition to decreased cell proliferation, terminal deoxynucleotidyl transferase-mediated dUTP nick end-labeling (TUNEL) and caspase assay revealed that apoptosis was increased in newly committed TCTP-disrupted cells as they migrated away from the ventricular zone. The mechanism may be that the phenotype from specific deletion of TCTP in neural progenitor cells is correlated with the decreased expression of cyclins D2, E2, Mcl-1, Bcl-xL, hax-1, and Octamer-binding transcription factor 4 (Oct4) in conditional knockout mice. Our results demonstrate that TCTP is a critical protein for cell survival during early neuronal and glial differentiation. Thus, enhanced neuronal loss and functional defect in Tuj1 and doublecortin-positive neurons mediated through increased apoptosis and decreased proliferation during central nervous system (CNS) development may contribute to the perinatal death of *TCTP* mutant mice.

Keywords: apoptosis; conditional knockout mice; development; Nestin-cre; neurogenesis; neuronal progenitor cells; perinatal death; proliferation; TCTP

1. Introduction

Translationally controlled tumor-associated protein (TCTP) is a highly conserved and abundantly expressed protein that has been implicated in both cell growth and the human acute allergic response. TCTP is widely expressed in many tissues and cell types [1], and its protein levels are highly regulated in response to a wide range of extracellular signals and cellular conditions [2]. TCTP has been shown to interact with tubulin [3], Na^+, K^+-ATPase [4], mammalian Plk [5], translation elongation factors eEF1A and eEF1Bbeta [6,7], TSAP6 [8], Mcl-1 [9], Bcl-XL [10], vitamin D3 receptor (VDR) [11], and p53 [12] for cell cycle regulation, protein synthesis, and antiapoptosis.

TCTP is also known as a histamine-releasing factor that stimulates the secretion of histamine [13]. In addition, decreased TCTP levels have been detected in certain areas of postmortem brains from Down syndrome and Alzheimer's patients [14]. In previous studies, we demonstrated that TCTP functions as an antiapoptotic protein required for mouse embryonic development and survival in systemic knockout mice [15]. Endoderm markers', such as Shh and PECAM, expression patterns were markedly reduced in embryonic day 9.5 (E9.5) $TCTP^{-/-}$ embryos, especially in the forebrain and midbrain anterior end. Furthermore, TCTP plays a very modest role in thymocyte development but is critical for peripheral T cell maintenance and TCR-mediated cell proliferation [16]. TCTP is also essential for developmental β-cell mass establishment and adaptation in response to insulin resistance [17]. Recent studies demonstrated that the interacting proteins of TCTP, such as Mcl-1 and Bcl-xL, play roles in the brain. Mcl-1 is a key regulator of apoptosis during the development of the central nervous system (CNS) [18]. Bcl-xL germline-deficient mice also undergo embryonic death at E13 and show increased apoptotic activity in the brain [19]. The conditional disruption of Bcl-xL in catecholaminergic neurons results in viable mice, with the catecholaminergic neuronal population reduced by one-third [20]. These studies suggest that TCTP may play an important role in CNS development. However, despite the importance of TCTP in the regulation of apoptosis, the impact of TCTP disruption during CNS development has not been investigated, except one recent paper that showed that TCTP may be implicated in the regulation of visual axon development of *Xenopus laevis* [21]. Therefore, the neuronal function of TCTP in the brain requires further investigation.

In the present work, we generated and characterized the phenotype of *TCTP* mutant mice and determined the possible mechanisms involved. We showed with a mouse model that TCTP is required for neural development in mammals. Deficiency of TCTP in neuronal and glial cell precursors resulted in decreased bromodeoxyuridine (BrdU) incorporation, increased widespread apoptosis, and disturbance of Tuj1-positive cell maturation, subsequently leading to perinatal death of TCTP mutant mice. Taken together, our results demonstrate that TCTP is required for the survival and differentiation of neuronal progenitor cells and is essential for cortical neurogenesis in development.

2. Materials and Methods

2.1. Generation of Conditional Knockout Mice, Breeding, and Genotyping

Mice harboring the floxed allele (f) of the *TCTP* gene were generated and genotyped as previously described [15]. Brain neuronal progenitor cell-specific TCTP conditional mutants were obtained by breeding *floxed TCTP* mice with *Nestin-cre* mice (B6.Cg-*Tg (Nes-cre)1Kln*/J, #003771) from Jackson Laboratory [22,23] to produce $Nestin^{Cre/+}$; $TCTP^{flox/+}$ (heterozygous, het) mice. The $Nestin^{Cre/+}$; $TCTP^{flox/+}$ mice were crossed with $Nestin^{Cre/+}$; $TCTP^{flox/+}$ mice or $TCTP^{f/f}$ alone mice to produce $Nestin^{Cre/+}$; $TCTP^{flox/flox}$ homologous conditional mutant mice (TCTP-cKO). $Nestin^{Cre/+}$; $TCTP^{+/+}$, or $TCTP^{f/f}$ alone mice were used as a control. Both *floxed TCTP* and *Nestin-Cre* mouse lines were generated in C57BL/6 and 129svj mixed background, and the mice used in this study were back-crossed to C57BL/6 for at least 8 generations. Double-heterozygous littermates ($Nestin^{Cre/+}$; $TCTP^{f/+}$) were also used as controls for some in vivo experiments. For embryonic time points, the time of plug identification was counted as postnatal day 0.5 (P0.5). Genotyping was performed by PCR using primers P1 (5′-TCTAGAAAAGTGGAGGCGGAGC-3′) and P5 (5′-GGTGACTACTGTGCTTT CGG TA-3′) for the wild-type (450 base pairs) and floxed (520 base pairs) alleles, and cre-sense (5′-TGCCACGACCAAGTGACAGC-3′) and cre-antisense (5′-CCTTAGCGCCGTAAAT CAATCG-3′) for the cre allele (580 base pairs). All animal studies were performed following the recommended procedures approved by the Institutional Animal Care and Use Committee of Tzu Chi University (PPL number: 97043) and conformed to the guidelines of Directive 2010/63/EU of the European Parliament on the protection of animals used for scientific purposes or the National Institutes of Health (NIH) guidelines.

2.2. Tissue Processing, Histological Analysis, and Immunohistochemistry

For histological analysis, whole brains were fixed with 4% paraformaldehyde and embedded in paraffin, sectioned, and stained with hematoxylin and eosin. TCTP expression pattern, BrdU incorporation, terminal deoxynucleotidyl transferase-mediated dUTP nick end-labeling (TUNEL) assay, caspase activation, and Oct4 expression were detected by immunohistochemistry with DAB staining. Immunofluorescence was performed by labeling with anti-TCTP (1:200; Abcam, Cambridge,

United Kingdom), anti-Tuj1 (1:100; Millipore, Burlington, MA, USA), anti-active caspase-3 (1:100; Abcam), anti-nestin (1:100; Cell Signaling, Danvers, MA, USA), and anti-MAP (1:100; Invitrogen, Carlsbad, CA, USA) antibodies. The retrieved sections were incubated with the primary antibody overnight at 4 °C, washed for 1 h in phosphate-buffered saline (PBS), and incubated with the secondary antibody for 1 h, then washed for 1 h in PBS. Sections were followed by DAB staining, and then counterstained with hematoxylin. For the immunofluorescence sections, visualization of staining was achieved using HiLyte Fluor 488- and HiLyte Fluor 555-conjugated secondary antibodies (1:200; AnaSpec, Fremont, CA, USA), then counterstained with 4', 6-diamidino-2-phenylindole (DAPI) for double or triple immunofluorescence staining. Immunofluorescence was visualized and captured using an Olympus BX43 upright fluorescent microscope equipped with a digital camera system (DP-72, Olympus, Tokyo, Japan) or a Zeiss LSM 510 META confocal imaging system (Carl Zeiss, Oberkochen, Germany).

2.3. RNA Isolation, cDNA Synthesis, and Quantitative Real-Time Reverse Transcriptase PCR

To analyze gene expression in the embryonic and postnatal stages of control and TCTP-cKO mice, RNA was isolated from 4 independent biological samples at stages E13.5, E16.5, P0.5, and P56 using TRIzol reagent (Invitrogen #15596018). Total RNA (1 μg) extracted from brain tissue was reverse-transcribed to synthesize cDNA using a ReverTra Ace-α- reverse transcription kit (Toyobo #F0937K, Toyobo, Osaka Prefecture, Japan) and oligo (dT) as a primer. Real-time PCR was performed using Power SYBR Green PCR Master Mix (Applied Biosystems #4368577, Applied Biosystems, Foster City, CA, USA) in an ABI 7300 Real-Time PCR System (Applied Biosystems). The specific primers used for the detection of genes are shown in Table 1. Each reaction was performed in duplicate, and dissociation curves were constructed to ensure that only a single product was amplified. The transcript expression level of the gene of interest was normalized to GAPDH mRNA as the internal control, and the data were expressed as a fold change over the control group.

2.4. Cell Proliferation Assay

Bromodeoxyuridine (5-bromo-2-deoxyuridine, BrdU) is a synthetic nucleoside that is an analogue of thymidine. BrdU can be incorporated into the newly synthesized DNA of replicating cells (during the S phase of the cell cycle), substituting for thymidine during DNA replication. For the labeling of cells in the S phase, pregnant mice (E13.5, E16.5, E17.5) obtained from timed mating were injected intraperitoneally with 10 mg BrdU (Invitrogen #000103) per 100 g of body weight. The brains dissected from E13.5, E16.5, and neonatal offspring (P0.5) were recovered in ice-cold PBS at pH 7.4 and fixed in 4% paraformaldehyde. Incorporation of modified nucleotide was detected by staining with anti-BrdU antibody (BU-20) as described previously [24].

2.5. TUNEL Assay

Cells in embryos undergoing apoptosis were analyzed by TUNEL assay using the In Situ Cell Death Detection Kit (Roche Diagnostics, Basel, Switzerland) according to the manufacturer's instructions. After TUNEL staining, the sections of brain were counterstained using hematoxylin.

2.6. Primary Neuronal Cultures

Primary cortical culture was prepared from fetal cortices of $Nestin^{Cre/+}$; $TCTP^{f/+}$ intercrossed with $TCTP^{f/f}$ mice at embryonic day 16.5 (E16.5) or postnatal day 0.5 as previously described [25]. Briefly, the fetal cortices were removed and dissected, followed by mechanical trituration in Hanks' balanced salt solution (GIBCO #14185, Thermo Fisher, Waltham, MA, USA) containing 2.5 U/mL dispase and 2 U/mL DNase. The supernatant that contained cortical neurons was filtrated through a 70-μm filter (BD Falcon #REF352350, New York, NY, USA) and transferred into a 15-mL autoclaved tube, and then immediately centrifuged at 1500× g for 10 min. The pellet containing neurons was resuspended in minimum essential medium (MEM) (GIBCO #12561) containing 10% heat-inactivated fetal bovine serum (FBS), 10 g/L glucose (Sigma #G7021, St. Louis, MI, USA), 0.176 g/L L-glutamine (GIBCO

#25030), 0.12 g/L sodium pyruvate (Sigma #p2256), 2.2 g/L sodium bicarbonate, 0.238 g/L HEPES (Sigma #H4034), and 10 mL/L 100× penicillin–streptomycin (BioWest #L0022, Les Ulis, France). Cells were seeded at a density of 2.5 × 10^5/well in 0.5 mL medium in a 24-well culture plate. The culture dishes were precoated with poly-D-lysine hydrobromide (50 µg/mL) (BD Bioscience #354210) for 2 h. The dishes were then washed twice with autoclaved deionized water. After 4 h, the MEM was replaced by Neurobasal medium (GIBCO #21103-049) supplemented with B27 (GIBCO #17504-044). Cells were incubated at 37 °C in a humidified atmosphere of 5% CO_2 and 95% air.

2.7. Cortical Progenitor Cultures and Immunofluorescence

Cortical progenitor cells were cultured as described previously [26]. Briefly, cortices were dissected from TCTP-cKO and littermate control embryos at E14.5–E15.5. Cortices were mechanically dissociated by trituration, and cell aggregates were plated on polyornithine-coated 4-well dishes and cultured in media containing Neurobasal medium (Invitrogen), 0.5 mM glutamine, 0.5 % penicillin–streptomycin, 1% N2 supplement (Invitrogen), 2% B27 supplement (Invitrogen), and 10 ng/mL NGF-2. On day 1, the medium was replaced with fresh medium. Immunofluorescence or immunohistochemistry experiments were performed 3 days after culture. Cultured cells were fixed in 4% paraformaldehyde for 20 min at room temperature and further processed for immunostaining. Cells were permeabilized with 0.1% Triton X-100, blocked for 1 h in 5% bovine serum albumin–5% goat serum, and incubated with primary antibodies, rabbit TCTP, and anti-nestin. After incubation overnight, cells were washed with PBS followed by 2 h of incubation with secondary antibodies, conjugated FITC, or rhodamine. Cells were counterstained for 30 s with DAPI for double immunofluorescence.

2.8. Cell Survival Assay and MTT Reduction Assay

Quantitative measurements of cortical progenitor cell survival were performed using the 3-[4, 5-dimethylthiazol-2-yl]-2, 5-duphenyltetrazolium bromide (MTT) reduction assay. The MTT reduction assay measures mitochondrial function as an index for cell survival. Cells were incubated in culture medium with MTT at a final concentration of 0.5 mg/mL MTT for 2 h; afterwards, the culture medium was replaced with 500 mL of dimethyl sulfoxide. Absorbance at 570 nm was determined by an ELISA reader.

2.9. Statistical Analysis of the Data

The experimental results are expressed as the mean and standard error of the mean (mean ± SEM) accompanied by the number of observations. Data were analyzed by Student's unpaired *t*-test, taking *p*-values of <0.05 as significant. Data obtained from 3 or more groups were compared by one-way ANOVA; *p*-values of <0.05 were considered statistically significant. Because the sample sizes involved in each experiment were different, they were included in the figures and/or figure legends.

3. Results

3.1. Expression Pattern of TCTP in the Mouse Brain during Development

We demonstrated that TCTP plays a critical role in survival at the mid-embryonic stage, but its role and functional mechanism in the regulation of neurogenesis at the early stage of brain development remains unknown. To address this question, we first examined whether TCTP is expressed in the developing brain and other populations of cells. TCTP RNA expression can be detected in the neural ectoderm and nervous system in mouse embryos at the E10.5 stage with in situ hybridization (Supplementary Figure S1), and the whole brain in E13.5, E16.5, P0.5, and P56 mice with qPCR (Figure 1A) with specific primers (Table 1), suggesting the presence of TCTP RNA from E13.5 to P10 involved in neural development. To gain insight into the TCTP function during brain development, we analyzed TCTP protein expression profiles in the CNS during development. As shown in Figure 1B, immunoblotting with a specific anti-TCTP antibody revealed that TCTP protein was highly expressed in whole brains from E13.5, E16.5, P0.5, and P10 but not mature mice (P56), as shown in the upper panel. The lower panel shows a similar TCTP protein expression pattern from E12.5 to adult. A widespread TCTP protein expression pattern was confirmed by immunohistochemistry analysis in whole-brain sections from mice at E13.5 (Figure 1C), E16.5 (Figure 1D), and postnatal day 0.5 (Figure 1E), including the striatum, olfactory bulb, cerebral cortex, hippocampus, cerebellum, and brain stem. The quantification data are summarized in Figure 1F. Furthermore, TCTP protein at E13.5 detected by double immunofluorescence staining was highly expressed in the ventricular zone (Figure 1G), where neural progenitor cells are concentrated. We found that TCTP expression in the ventricular zone was dramatically decreased to a low level from E16.5 to P0.5 but increased in the cortical layer. The abundant expression pattern of TCTP in the brain suggests that it may play an important role in CNS development. To further investigate the role of TCTP in nervous system development, we disrupted TCTP expression with conditional knockout mice, specifically in the neural progenitor cells of the brain.

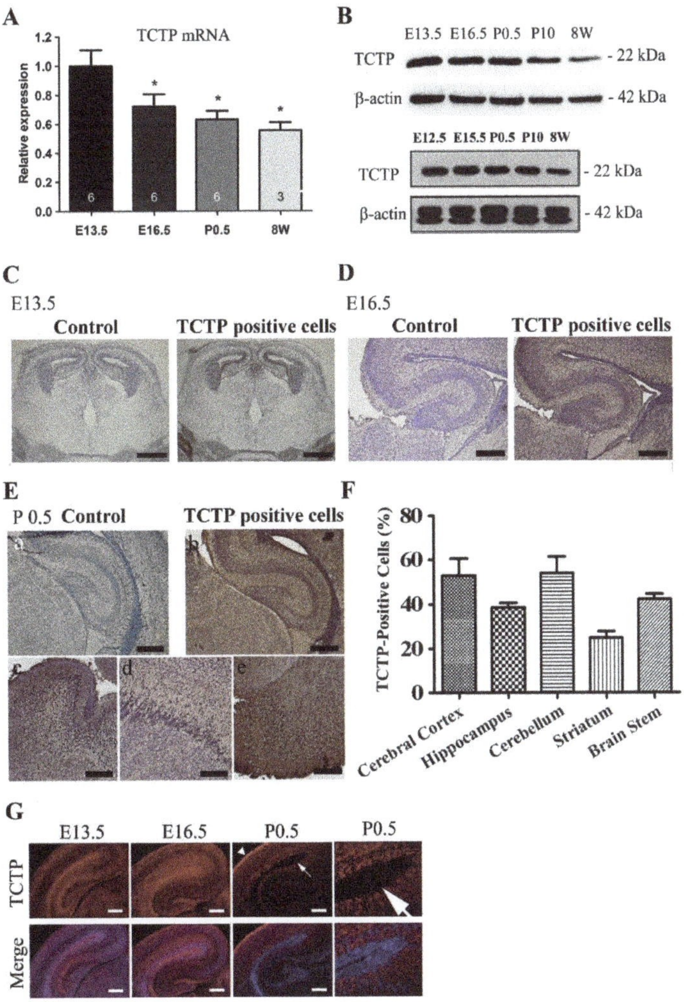

Figure 1. Expression pattern of translationally controlled tumor-associated protein (TCTP) from the embryonic to adult stage in the brain. (**A**) Relative TCTP mRNA expression at different stages in the

whole brain from wild-type mice was estimated by qRT-PCR (data presented as mean ± SEM, * $p < 0.05$ compared with embryonic day 13.5 (E13.5), one-way ANOVA, n = 3–6 per group). (**B**) TCTP protein in the whole brain from control mice (*TCTP$^{flox/flox}$*) at E12.5, 13.5, E15.5, 16.5, and postnatal day 0.5, 10, and 56 was estimated by Western blotting. (**C–E**) TCTP protein expression pattern in whole-brain sagittal sections from wild-type mice at E13.5 and coronal sections at E16.5 and P0.5, including the (a) cerebral cortex, (b) hippocampus, (c) cerebellum, (d) striatum, and (e) brain stem, detected by immunohistochemistry with DAB staining (brown stain indicates TCTP-positive cells) (n = 3 for each time point). (**F**) Summarized quantification of TCTP protein expression in (**E**). (**G**) Representative TCTP (red) and 4′,6-diamidino-2-phenylindole (DAPI)/TCTP double immunofluorescence staining of brain sections from C57BL/6 mice at E13.5, E16.5, and P0.5; n = 3. The arrow indicates the ventricular zone area; the arrowhead indicates the cortical layer. Scale bars: C, 600 μm; D, 200 μm; E: a, c, d, e, 100 μm, b, 200 μm; G, 200 μm.

3.2. Generation of Conditional TCTP Knockout Mice

To circumvent the embryonic lethality of complete disruption of the *TCTP* gene [15] and to examine the role of TCTP in CNS development, we used the Cre/loxP gene targeting system to generate neuronal progenitor cell-specific TCTP conditional mutants, which were obtained by breeding *TCTP$^{f/f}$* mice with *Nestin-cre* transgenic mice from Jackson Lab [22,23] to produce TCTP-cKO mice (Figure 2A). The *Nestin-Cre* transgenic mice expressed Cre recombinase under the control of the rat *Nestin* promoter. The *Nestin-cre*-mediated recombination was first present in precursor cells of neurons and glia around embryonic day 11 and was detected in the cortical wall, cortical layers, ventricular zone (VZ), subventricular zone (SVZ), and intermediate zone (IZ) of the telencephalon and spinal cord of the growing population of postmitotic neurons in the developing central nervous system [27]. The generation and characterization studies of TCTP-cKO conditional mutants could help us to selectively assess the role of TCTP in neural precursor cells during central nervous system development.

Genotype analysis of progeny at P10 from *Nestin$^{Cre/+}$; TCTP$^{f/+}$* mice crossed with *TCTP$^{f/f}$* revealed that no homozygous offspring were found, suggesting embryonic or neonatal death of TCTP-cKO mice. Therefore, we checked the embryo variability for TCTP knockout mice. Homozygous offspring could be found at E9.5, E10.5, E13.5, E14.5, and E16.5. The TCTP disruption genotype *Nestin$^{Cre/+}$; TCTP$^{f/f}$* was confirmed by PCR in newborn mice, postnatal day 0.5 (Figure 2B). The four offspring genotypes (*Nestin$^{+/+}$; TCTP$^{f/f}$*, *Nestin$^{+/+}$; TCTP$^{f/+}$*, *Nestin$^{Cre/+}$; TCTP$^{f/+}$*, *Nestin$^{Cre/+}$; TCTP$^{f/f}$*) were identified in an expected Mendelian ratio (1:1:1:1). The *Nestin-cre*-derived deletion of TCTP was further demonstrated by TCTP mRNA level with qRT-PCR (Figure 2C), and protein expression with immunoblotting (Figure 2B, lower panel) and immunohistochemistry (Figure 2D). The sections at E16.5 were also detected by immunofluorescence double labeled with antibodies to TCTP to identify the *TCTP* deletion and antibodies against *Nestin* to identify neural precursor cells (Figure 2E). Mice heterozygous for the deleted allele of *TCTP* (*Nestin$^{Cre/+}$; TCTP$^{f/+}$*) were viable, fertile, and not morphologically different from wild-type (*Nestin$^{Cre/+}$; TCTP$^{+/+}$*) littermates.

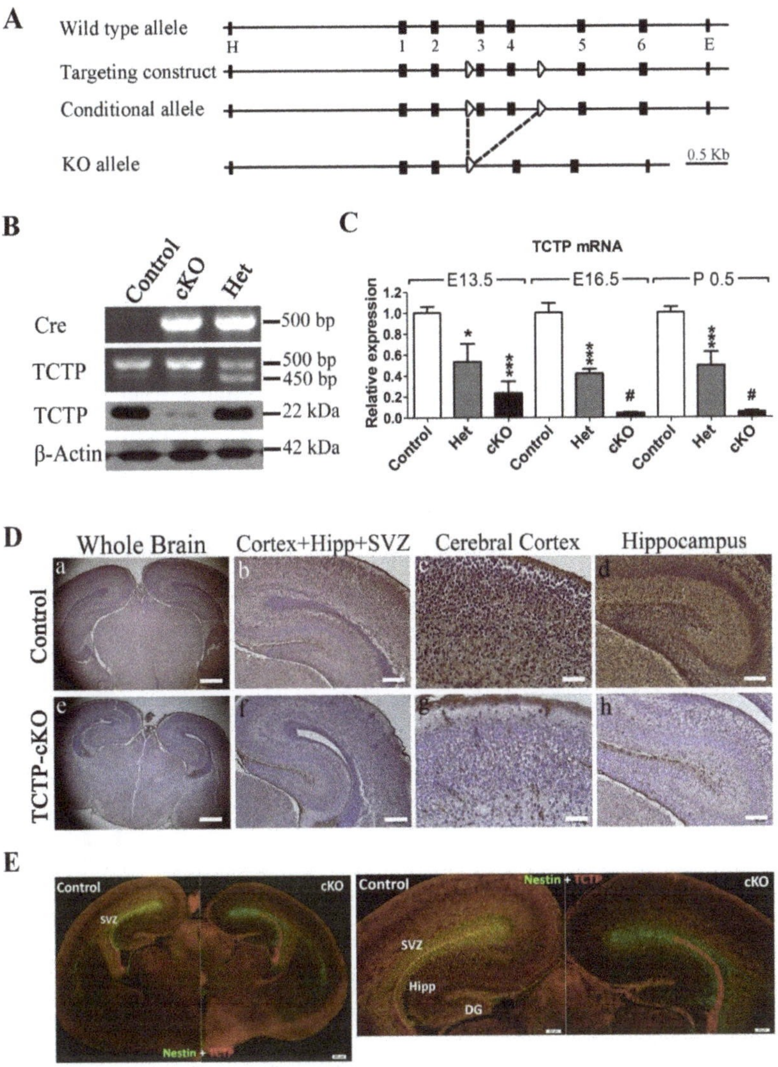

Figure 2. Generation of conditional TCTP knockout mouse model. (**A**) Diagram of the targeting strategy for the generation of the conditional TCTP allele. The targeting construct contains *loxP* sites (arrowheads) flanking exons 3–4 and restriction sites EcoRI (E) and HindIII (H). (**B**) Genotyping with PCR, top panel, and Western blotting of TCTP expression, bottom panel, from $Nestin^{+/+}$; $TCTP^{flox/flox}$ (Control), $Nestin^{Cre/+}$; $TCTP^{flox/flox}$ (heterozygous), and $Nestin^{Cre/+}$; $TCTP^{flox/+}$ (cKO) mice at P0.5. (**C**) qRT-PCR of relative TCTP mRNA expression in the whole brain from control, heterozygous, and conditional knockout mice at different stages was estimated (data presented as mean ± SEM, * $p < 0.05$, *** $p < 0.001$ compared with control, # < 0.05 compared with E13.5 cKO, one-way ANOVA, n = 3–11 per group). (**D**) Immunohistochemistry of TCTP protein from control and TCTP-cKO mice at P0.5. DG, dentate gyrus; Hipp, hippocampus; SVZ, subventricular zone. Scale bars: a, e, 400 μm; b, f, 200 μm; c, g, 40 μm; d, h, 100 μm. (**E**) Sections at E16.5 were also detected by double-labeled immunofluorescence with antibodies against TCTP to identify the deletion and Nestin to identify neural precursor cells. A representative image is shown for three independent experiments. Scale bars: 100 μm for right panel, 200 μm for left panel.

3.3. Loss of TCTP in Neuronal Progenitor Cells Resulted in Early Neonatal Death

Compared with littermate controls, all the surviving TCTP-cKO mutant pups at P0.5 were smaller in body size (Figure 3A) and did not suckle well (Figure 3B, indicated by arrow), as demonstrated by the lack of milk in their stomachs. Abnormal behavior, particularly ataxia, was observed in cKO pups. Moreover, body weight, brain size, and brain weight were reduced in TCTP mutant pups examined at P0.5 when compared with littermate controls (Figure 3C,E,F). We found that the TCTP mutant mice exhibited early postnatal lethality phenotype at P1.5 to P2.5 (Figure 3D, Table 1). Mice heterozygous for the deleted allele of *TCTP* ($Nestin^{Cre/+}$; $TCTP^{f/+}$) were viable, fertile, and not morphologically different from wild-type ($Nestin^{Cre/+}$; $TCTP^{+/+}$) littermates. The perinatal death of TCTP-cKO mice might be caused by abnormal CNS development and malnutrition from a lack of milk in the stomach.

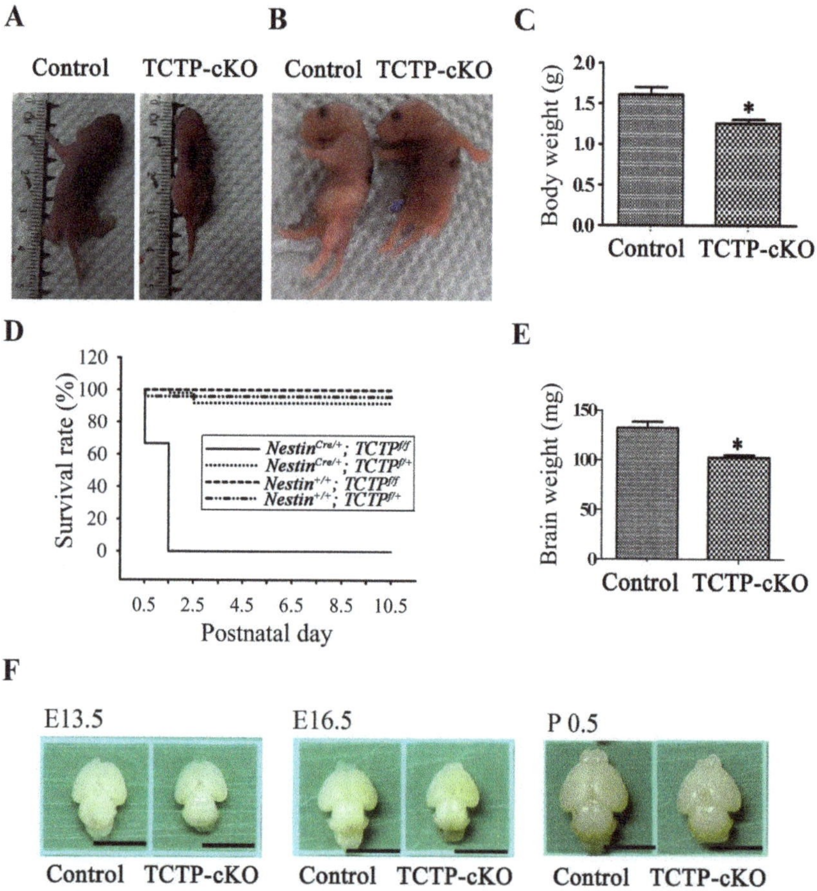

Figure 3. Phenotypes and survival rates of TCTP-cKO mice. (**A**) Body size of control and TCTP-cKO mice at P0.5. (**B**) Milk in stomachs of control and cKO mice at P0.5. (**C**) Body weight of control and cKO mice at P0.5 (data presented as mean ± SEM, * $p < 0.05$, Student's t test, $n = 3$ per group). (**D**) Survival rate of offspring from $Nestin^{Cre/+}$; $TCTP^{f/+}$ mice crossed with $TCTP^{f/f}$ mice during development from P0.5 to P10.5. (**E**) Brain weight of control and cKO mice at P0.5 (data presented as mean ± SEM, * $p < 0.05$, Student's t test, $n = 4$ per group). (**F**) Morphology of brains from control and cKO mice at E13.5, E16.5, and P0.5. Scale bar: F, 0.5 cm.

Table 1. Genotype analysis of offspring from *Nestin*$^{Cre/+}$; *TCTP*$^{f/+}$ mice crossed with *TCTP*$^{f/f}$ mice.

Stage	Embryonic Time Point							
	E9.5~E10.5	E12.5	E14.5	E16.5	P0.5	P1.5	P3.5	P10.5
Litter	6	4	5	4	11	5	8	11
Total number of Offspring	37	22	31	33	73	38	48	71
Nestin$^{Cre/+}$; TCTP$^{f/f}$	10 (27%)	5 (23%)	8 (27%)	7 (21%)	25 (34%)	7 (18%)	0 (0%)	0 (0%)
Nestin$^{Cre/+}$; TCTP$^{f/+}$	12 (32%)	5 (23%)	10 (32%)	9 (27%)	25 (34%)	16 (42%)	16 (33%)	28 (39%)
Nestin$^{+/+}$; TCTP$^{f/f}$	5 (14%)	8 (36%)	6 (19%)	9 (27%)	14 (19%)	6 (16%)	19 (40%)	22 (31%)
Nestin$^{+/+}$; TCTP$^{f/w}$	10 (27%)	4 (18%)	6 (19%)	8 (24%)	9 (12%)	9 (24%)	13 (27%)	21 (30%)

3.4. TCTP is Required for Cortical Neurogenesis

The perinatal death of TCTP-cKO mice prompted us to further detect the brain morphology of cKO mice compared with littermate controls by hematoxylin and eosin Y staining. To assess the role of TCTP in brain development, brains were collected at E16.5, around mid-neurogenesis, and P0.5. This stage is characterized by a large population of Nestin-expressing progenitors within the VZ and subventricular zone (SVZ), early committed neuroblasts within the SVZ/intermediate zone (IZ), and a growing population of postmitotic neurons in the developing cortical plate (CP). Therefore, we examined the cortical plate, hippocampus, and lamination of SVZ in the cerebral cortex in TCTP-deficient brains compared with littermates (Figure 4A,B). The cortical plate and hippocampus region of the TCTP-cKO mouse brain was significantly decreased as expected. In contrast, the VZ/SVZ region in the cKO mouse brain was significantly increased. Enlargement of the VZ/SVZ region in TCTP-conditional null mice indicated that perhaps some progenitor cells cannot fully migrate and commit to differentiation or are unable to exit the cell cycle. TCTP-cKO mice at P0.5 also exhibited a size reduction in the cortical plate and hippocampus but broad and lower levels of staining in the proliferative SVZ and VZ compared with littermate controls (Figure 4C,D). The structural abnormality in the mutant brain at E16.5 and P0.5 suggests that the selective deletion of TCTP in neural progenitor cells significantly impaired the development of the cerebral cortex surrounding the lateral ventricle (Figure 4).

Furthermore, Tuj1 was expressed not only in newly committed immature postmitotic neurons but also in differentiated neurons and in some mitotically active neuronal precursors [28]. Therefore, the brain sections and isolated primary cortical neurons from control and TCTP conditional knockout mice at E16.5 or P0.5 were prepared for detection by double immunofluorescence staining with antibody against TCTP and Tuj1. We found that the expression of Tuj1 was reduced in the cortex and hippocampus of the brain sections (Figure 5A,B) and primary cortical neurons (Figure 5C,D) from TCTP-cKO mice at E16.5 and P0.5 by immunofluorescence analysis. The results exhibited decreased neurogenesis in the TCTP cKO mice compared with control mice (Figure 5A–D). To further investigate the impact of TCTP deletion on neurogenesis, we examined several cell-type-specific markers in the cerebral cortex, hippocampus, VZ, and SVZ. Nestin protein is expressed at high levels in cortical radial glia/neural progenitor cells (NPCs) [29]. In contrast, doublecortin (DCX) expression is low in NPCs but upregulated in postmitotic neurons [30]. Nestin and DCX are thus routinely used as markers to distinguish these mutually exclusive cell types. DCX, a marker for newly migrating populations, could affect neuronal migration by regulating the organization and stability of microtubules [31]. Compared to littermate controls, the brains of cKO mutant mice exhibited a significant reduction of neurogenesis marker DCX-positive neurons (Figure 5E,F) by immunohistochemistry analysis. Moreover, an in vitro study of primary cell culture indicated that high expression of TCTP was observed in both proliferating neural precursors (Nestin+) and mature neurons (MAP2+) during brain development. Some TCTP-negative cells from mutant mice expressed Nestin and survived at first (Supplementary Figure S2A). In contrast, most surviving cells cultured from mutant mice expressed TCTP and MAP but not TCTP-negative cells (Supplementary Figure S2B,C). These results indicate that TCTP is involved in neurogenesis in the embryonic and early postnatal stages. Specifically, *Nestin-cre*-derived TCTP disruption resulted in decreased Tuj1-positive newly committed neurons, DCX, and Nestin-positive neural progenitors, indicating that TCTP is required for the survival of these cell populations. Furthermore, we also demonstrated that a deficiency of TCTP possibly delayed or impaired neuronal proliferation and cell cycle progression.

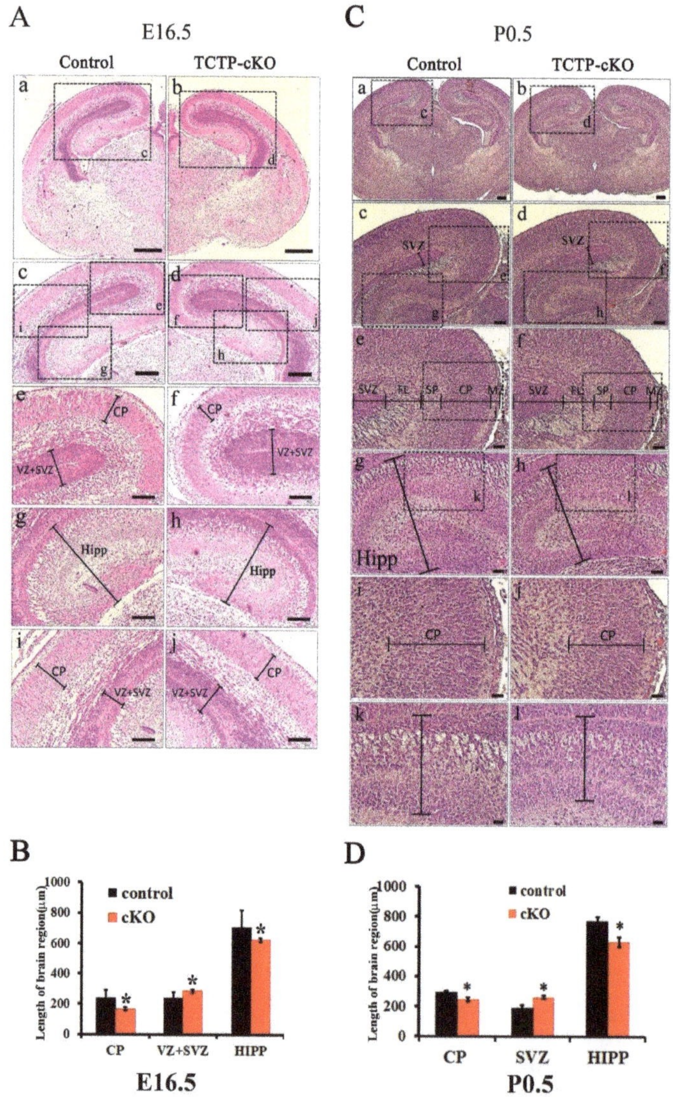

Figure 4. Histological analysis of E16.5 embryos and P0.5 pups from control and cKO mice. (**A**) Control (*Nestin*$^{Cre/+}$; *TCTP*$^{w/w}$; a, c, e, g, and i) and cKO (*Nestin*$^{Cre/+}$; *TCTP*$^{f/f}$; b, d, f, h, and j) brain sections from E16.5 embryos were stained with hematoxylin and eosin. (**B**) Bar graph summarizing quantification of the sizes of brain regions in (**A**). (**C**) Control (a, c, e, g, i, and k) and cKO (b, d, f, h, j, and l) brain sections from P0.5 pups were stained with hematoxylin and eosin. (**D**) Bar graph summarizing the quantification of the sizes of brain regions in (**C**), (data presented as mean ± SEM, * $p < 0.05$, Student's *t* test, $n = 3$ per group). A representative image is shown for three independent experiments. MZ, marginal zone; SP, subplate. Scale bars on panel A: a, b, 400 µm; c, d, 200 µm; e–j, 100 µm. Scale bars on panel B: a, b, 400 µm; c, d, 200 µm; e–h, 100 µm; i–l, 40 µm.

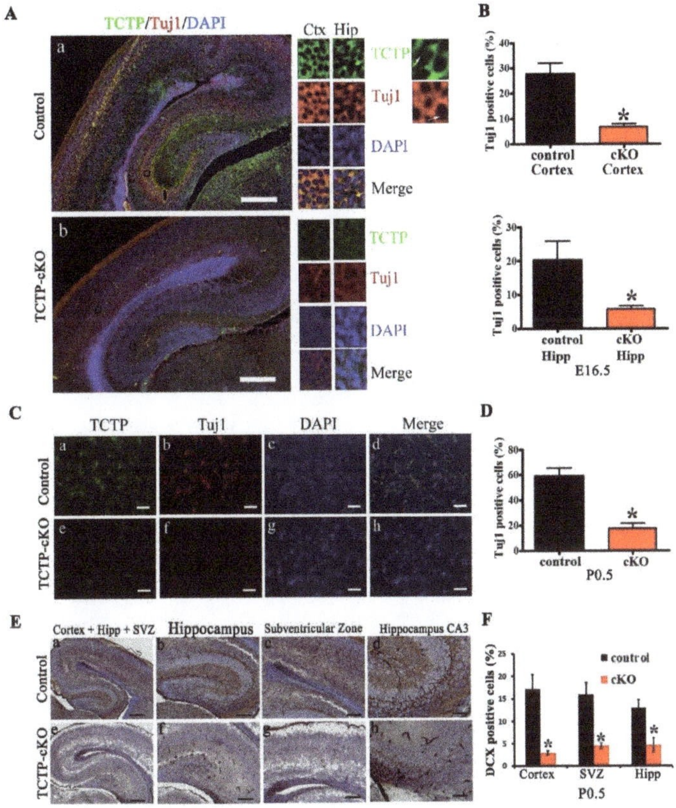

Figure 5. Central nervous system (CNS) neurogenesis of TCTP-cKO and control mice was detected. (**A**) Double immunofluorescence staining detected TCTP (green) and Tuj1 (red) of brain sections at E16.5. Bottom panel shows magnification of panels a and b. (**B**) Bar graph summarizing the quantification of Tuj1-positive cells in (**A**). (**C**) TCTP (green) and Tuj1 (red) of primary cultured cortical neurons from control and TCTP-cKO mice at P0.5 were detected. (**D**) Bar graph summarizing the quantification of Tuj1-positive cells in (**C**). (**E**) The neurogenesis marker doublecortin (DCX) was detected in TCTP mutant brains from control and TCTP-cKO mice at P0.5. (**F**) Bar graph summarizing the quantification of DCX-positive cells in (**E**). (Data presented as mean ± SEM, * $p < 0.05$, Student's t test, $n = 3$ per group.) Scale bars: A, 200 μm, B, 40 μm, C, a, e, 200 μm; b–d, f–h, 100 μm. A representative picture is shown for three independent experiments.

3.5. Decreased Cell Proliferation in Nestin-Cre-Derived TCTP-Deficient Mouse Brain

We next examined whether the cellular proliferation and apoptosis involved the phenotype of the TCTP-cKO mutant mice. To address this issue, BrdU incorporation and TUNEL assays were determined. We examined whether loss of TCTP causes the defect in brain development and neuronal cell proliferation. The proportion of cells in the S phase of the cell cycle was determined by immunohistochemistry. High levels of BrdU expression were detected in the hippocampus, dentate gyrus, marginal zone, and subventricular zone of the cerebral cortex in the littermate control mice at P0.5; however, BrdU-positive cells were reduced in the TCTP mutant mice (Figure 6A). The quantification data at P0.5 are summarized in Figure 6B. A significant reduction of BrdU incorporation was also found in the cKO mice at E16.5 (Figure 6C), especially in the corpus callosum, but not at E13.5 (Figure 6D). The quantification data at E16.5 and E13.5 are summarized in Figure 6E. These results are consistent with the *Nestin-cre* activity, which was high in the preplate and low in the ventricular zone (VZ) at

E12.5, and high in the cortical plate (CP) and intermediate zone (IZ) and low in the VZ and SVZ at E14.5. Then, the scope of recombination expanded to neural stem cells (NSCs) and NPCs increased dramatically during E14.5 to E17.5 [32]. To determine whether decreased proliferation was correlated with perturbation of cell cycle progression, immunoblotting analysis was carried out to compare the expression levels of cyclins D2 and E2 between the *Nestin*$^{Cre/+}$; *TCTP*$^{f/f}$ pups and littermate controls. We found that cyclin D2 and E2 expression was significantly suppressed in the TCTP-deficient brain (Figure 6F,G), which may delay or impair cell cycle progression and neuronal proliferation. These results were consistent with the BrdU incorporation assay and DCX immunohistochemistry staining. The reduced levels of cyclins D2 and E2 in *Nestin*$^{Cre/+}$; *TCTP*$^{f/f}$ pups was also consistent with what we demonstrated: That mice with systemic disruption of TCTP exhibited decreased expression of cyclin D2 and E2 in E9.5 TCTP$^{-/-}$ embryos [15]. These results indicate that TCTP is required for cortical neurogenesis.

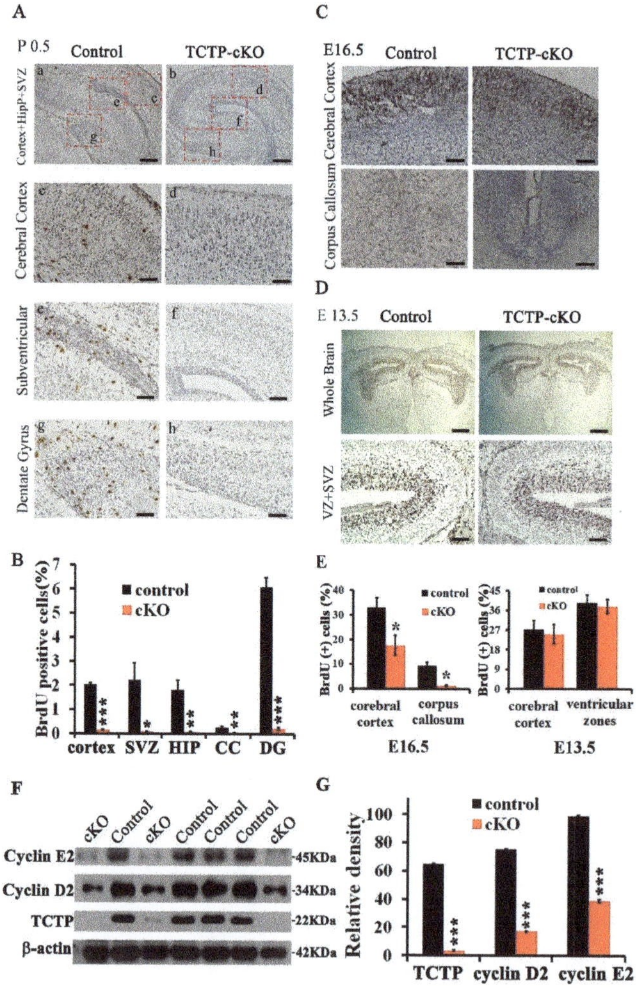

Figure 6. Cell proliferation and cyclin protein expression detected in brain sections and protein lysates from control and cKO mice. (**A**) Bromodeoxyuridine (BrdU) incorporation assay of control and TCTP-cKO mice was estimated at P0.5. (**B**) Bar graph summarizing the quantification of BrdU-positive

(+) cells in (**A**). (**C,D**) BrdU incorporation in brain sections from control and TCTP-cKO mice was also detected at E16.5 and E13.5 (**E**) Bar graph summarizing the quantification of BrdU-positive cells in (**C,D**). (Values are mean ± SEM, Student's *t* test, * $p < 0.05$ compared with control group; $n = 3$). CC. corpus callosum; DG, dentate gyrus; HIP, hippocampus; SVZ, subventricular zone. Scale bars: A: a, b, 200 μm; c, d, e, g, h, 40 μm; f, 100 μm. C: 40 μm; D: top panel, 400 μm; bottom panel, 40 μm. (**F**) Western blotting of cyclins D2 and E2 and TCTP in control and TCTP-cKO mice at P0.5. (**G**) Bar graph summarizing quantification of Western blotting relative density in (**E**). Relative density is presented as the ratio of band intensity of target protein to internal control-actin. Values are mean ± SEM. Data were statistically analyzed by student's *t* test; * $p < 0.05$, ** $p < 0.01$, *** $p < 0.001$ compared with control group; $n = 3$).

3.6. Disruption of TCTP-Induced Cell Apoptosis and Cell Autonomous Behavior

Next, we determined whether cell-specific disruption of TCTP expression leading to increased apoptotic cell death may contribute to the decrease in different kinds of neurons and neonatal death. To address this hypothesis, TUNEL assay was conducted at postnatal day 0.5. TUNEL staining was performed on coronal sections of the brain, including the hippocampus and corpus callosum. Increased TUNEL-positive cells (brown staining) were detected in *Nestin-cre*-derived TCTP-deficient brain compared with littermate controls (Figure 7A). The apoptotic cells were significantly increased in the cerebral cortex, subventricular zone, hippocampus, and corpus callosum in TCTP-cKO mutant compared with controls at P0.5 (Figure 7A,B) and E16.5 (Figure 7C) but not E13.5 (Figure 7D). The quantification data at E16.5 and E13.5 are summarized in Figure 7E.

We further evaluated whether the observed neuronal cell loss surrounding the mouse lateral ventricle resulted from apoptosis with another apoptosis marker. Coronal sections at P0.5 from *Nestin-cre*-mediated TCTP mutant brains and from control littermates were detected by immunohistochemistry with antibody agonist active caspase 3 (AC3), a hallmark of classical apoptotic cell death [18]. Numerous active caspase 3-positive cells were present throughout the cortical plate, VZ/SVZ, CP, and hippocampus of TCTP mutant mice, whereas only an occasional apoptotic cell was observed in littermate controls (Figure 8A,C). These data were consistent with the decrease of Tuj1- and DCX-positive neurons (Figure 5A,C,E) in mutant mice, suggesting that the increased apoptosis also involved the phenotype of the TCTP-cKO mutant mice. Immunohistochemistry with primary neuronal cultures was used to clarify whether the cell death of TCTP conditional knockout mice was cell autonomous or nonautonomous. The primary neuronal cultures from the TCTP-cKO mutant mice at E16.5 showed enhanced apoptotic cell death compared with control mice (Figure 8B,D). TCTP-negative cells (arrowhead), but not TCTP-positive cells (arrow), exhibited active caspase-3 positive signal (Figure 8B). This massive cell death by apoptosis observed in brain sections in vivo and TCTP-negative cells in vitro indicated that the cell death from *Nestin-cre*-derived TCTP disruption specifically in neuronal progenitor cells was caused by a cell autonomous mechanism. Again, TCTP is required for cortical neurogenesis.

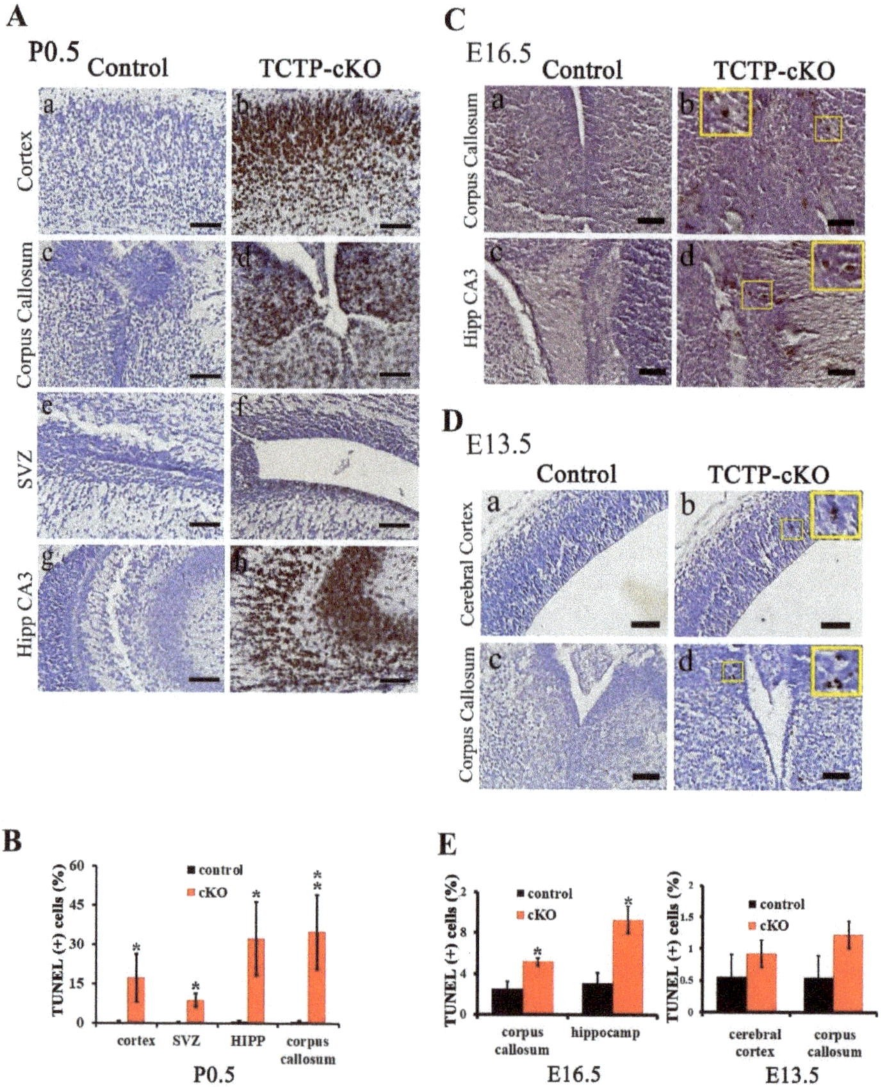

Figure 7. Cell apoptosis was detected in brain sections from control and cKO mice. (**A**) TUNEL staining was performed on coronal sections of the brain at P0.5, including the (a,b) hippocampus, (c,d) corpus callosum, (e,f) subventricular zone, and (g,h) hippocampus CA3. (**B**) Bar graph summarizing the quantification of TUNEL-positive (+) cells in (**A**). (Values are mean ± SEM, Student's *t* test, * $p < 0.05$, ** $p < 0.01$ compared with control group; $n = 3$). (**C**,**D**) Apoptosis of brain sections from TCTP-cKO and control mice at E16.5 ($n = 3$) and E13.5 ($n = 4$) was also detected by TUNEL staining assay. Boxes indicate magnification. (**E**) Bar graph summarizing quantification of TUNEL-positive cells in (**C**,**D**). (Values are mean ± SEM, Student's *t* test, * $p < 0.05$ compared with control group). Hipp, hippocampus; SVZ, subventricular zone. Scale bar: A, C, 40 μm; D, 100 μm.

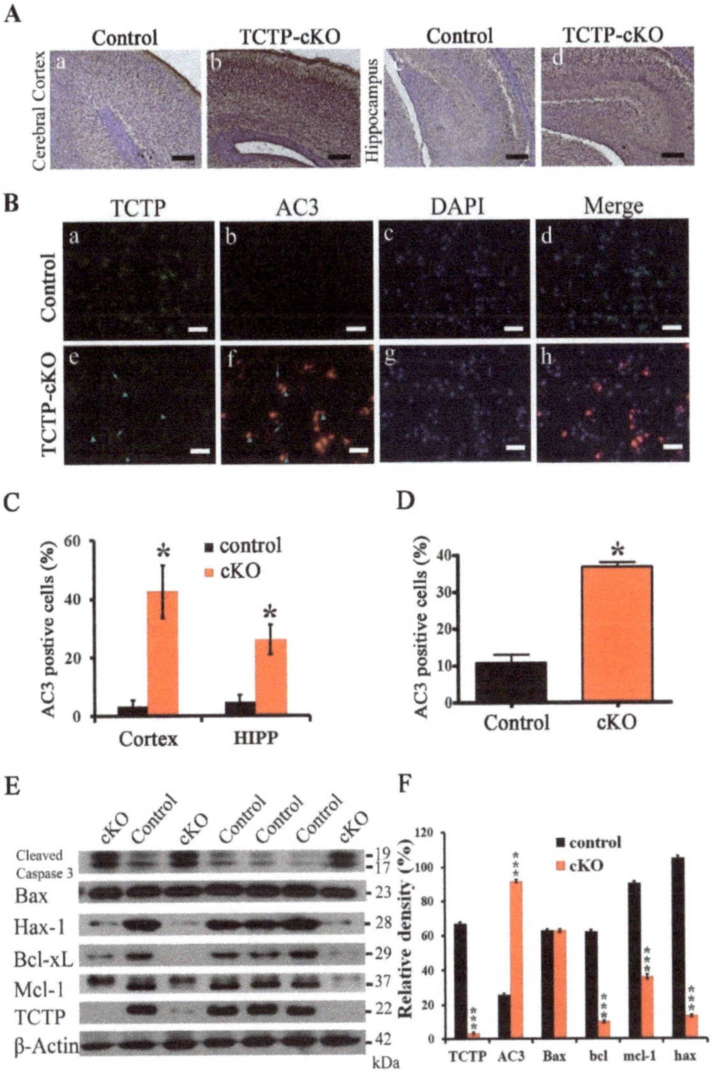

Figure 8. Caspase cleavage apoptosis was detected by immunohistochemistry, immunofluorescence, and Western blot. Apoptotic and antiapoptotic related protein levels were detected by Western blot. (**A**) Cerebral cortex and hippocampus of control and TCTP-cKO mice were stained for active caspase-3 (AC3) by immunohistochemistry at P0.5. (**B**) Double immunofluorescence staining for TCTP (green) and AC3 (red) of primary neuronal cell from control and TCTP-cKO mice at E16.5. The arrow indicates TCTP-positive cells, and the arrowhead indicates TCTP-negative cells. (**C**) Bar graph summarizing the quantification of active caspase-3 staining in (**A**). (**D**) Bar graph summarizing the quantification of active caspase-3 staining in (**B**). (Data presented as mean ± SEM, * $p \leq 0.05$, Student's t test, $n = 3$ per group). (**E**) Protein levels of cleaved caspase-3, bax, hax-1, bcl-xl, mcl-1, and TCTP in control and TCTP-cKO mice were measured by Western blotting at P0.5. (**F**) Bar graph summarizing quantification of Western blot in (**E**). Relative density is presented as the ratio of band intensity of target protein to the internal control, β-actin. (Data presented as mean ± SEM, *** $p < 0.001$, Student's t test, $n = 3$ per group). Scale bars: A, 100 μm; B, 40 μm.

Apoptotic cell death in the brain of TCTP-cKO mutant mice was also confirmed with the increased cleaved caspase-3 by Western blot. Among the interacting proteins of TCTP, we first investigated the expression of mMcl-1 [9] and Bcl-xL [10], both antiapoptotic members of the Bcl-2 family. TCTP protects from apoptotic cell death by antagonizing bax function in cooperation with mMcl-1 and Bcl-xL [33]. The *Nestin-cre*-driven TCTP knockout mice exhibited decreased expression levels of mMcl-1 and Bcl-xL but not bax protein (Figure 8E,F) compared with littermate controls. The disruption of TCTP caused Bcl-xL protein degradation (Figure 8E,F) but not decreased DNA transcription.

Antiapoptotic protein hax-1 was also reduced in mutant pups. These results demonstrate that TCTP plays a critical role in the developing nervous system when using TCTP-cKO mice. The increased apoptosis and decreased proliferation may contribute to early neonatal death. The phenotype of increased neuronal loss through apoptotic cell death in the brain of TCTP-cKO knockout mice might be attributed to the decreased cyclin D2, Mcl-1, Bcl-xL, and hax-1 expression. Therefore, TCTP is essential for the maintenance of neural precursor cell survival and differentiation during CNS development.

3.7. Conditional Deletion of TCTP in Neuron Progenitor Cells Resulted in Decreased Cell Survival and Suppression of Transcription Factor Oct4 Expression in Cortical Progenitor Cultures

To further confirm whether neural precursor cell death observed in vivo and in vitro is caused by a cell autonomous or nonautonomous mechanism, we cultured primary cortical progenitor cells and measured their survival in vitro as an indicator of cell survival using MTT assay. Cortical progenitor cell cultures from *Nestin-cre*-mediated TCTP mutant embryos at E15.5 did not exhibit significant differences in cell aggregates, dendrite outgrowth, and differentiation at 24 h compared with littermate controls in vitro. After 3 days of culture, decreased cell viability and cell numbers were observed in TCTP-deficient cells (Figure 9A). Quantification of total cells per aggregation revealed a significant four-fold reduction in TCTP-deficient cultures, indicating that the cell death from *Nestin-cre*-derived TCTP disruption is caused by a cell autonomous mechanism.

Niwa et al. showed that a critical amount of octamer-binding transcription factor 3/4 (Oct3/4) is required to sustain stem cell self-renewal, and up- or downregulation induced divergent developmental programs [34]. In addition, a previous study showed that Tpt1 activates transcription through binding to the regulatory region of the mouse *Oct4* gene in transplanted somatic nuclei in *Xenopus laevis* oocytes [35]. To examine whether the self-renewal of TCTP mutant mouse cortical progenitor cells may encounter a problem in undifferentiated embryonic stem cells, TCTP-modulated Oct4 expression was examined. Cortical progenitor cells were cultured from mouse brain at E15.5 and stained with antibody for Oct4 after 3 days of culture. Oct4-positive cells were detected in control groups, but a decrease was observed in the *Nestin-cre*-derived TCTP-deficient cell culture (Figure 9B). Moreover, we found significantly fewer cells and failure of differentiation of neuron dendrites in cortical progenitor cell culture. These results indicated that the loss of TCTP decreased cell numbers, suppressed transcription factor Oct4 expression, and disrupted cell differentiation in cortical progenitor cells cultured from *TCTP* mutant. The decreased Oct4 protein expression of TCTP-cKO mice was further confirmed in vivo by immunohistochemistry at E13.5 (Figure 9C) and P0.5 (Figure 9D,E) and Western blotting as early at E12.5 (Figure 9F) compared with littermate controls. Furthermore, we also studied the mRNA expression of Oct4 and cell survival-related genes with specific primers (Table 2). A significant decrease in Oct4 and Mcl-1 but not hax-1 and Bcl-xL mRNA in TCTP-cKO mice relative to control mice was observed (Figure 10). We expect that cells of cKO mice are needed for higher antiapoptotic activity, such as Bcl-xL and mcl-1, by the enhancement of mRNA expression levels. Bcl-xL protein was decreased, but Bcl-xL mRNA was increased in cKO mice at P0.5 (Figures 8C and 10E). On the other hand, mcl-1 was decreased in both protein and mRNA levels at P0.5 (Figures 8C and 10A).

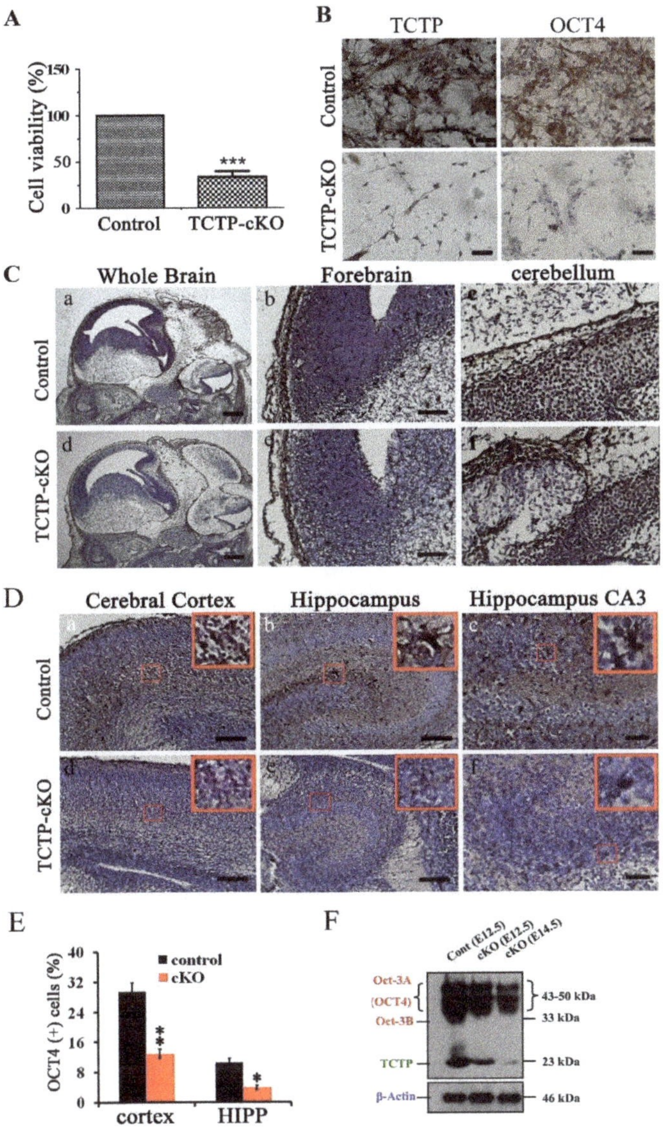

Figure 9. Cell viability and octamer-binding transcription factor 4 (Oct4) protein expression were detected in primary cortical progenitor cultures and brain sections. (**A**) Cell viability and cell number of neuron progenitor cell cultures from TCTP-deficient mice and littermate controls at E15.5 were detected by MTT assay. (*** $p < 0.001$, Student's *t* test, $n = 3$ per group). (**B**) TCTP and Oct4 expression was detected by immunohistochemistry with antibody followed by DAB staining and then counterstaining with hematoxylin. (**C**,**D**) Oct4 expression of control and TCTP-cKO mice at E13.5 and P0.5 was detected by immunohistochemistry. A representative picture is shown for two independent experiments. (**E**) Bar graph summarizing quantification of Oct4-positive (+) cells in (**D**), (Data presented as mean ± SEM, * $p < 0.05$, ** $p < 0.01$, Student's *t* test, $n = 3$ per group). (**F**) Protein level of Oct4 in control and TCTP-cKO mice was measured by Western blotting. Scale bar: A, 40 μm; C: a, d, 400 μm, b, c, e, f, 100 μm; D: a, b, d, e, 100 μm, c, f, 40 μm.

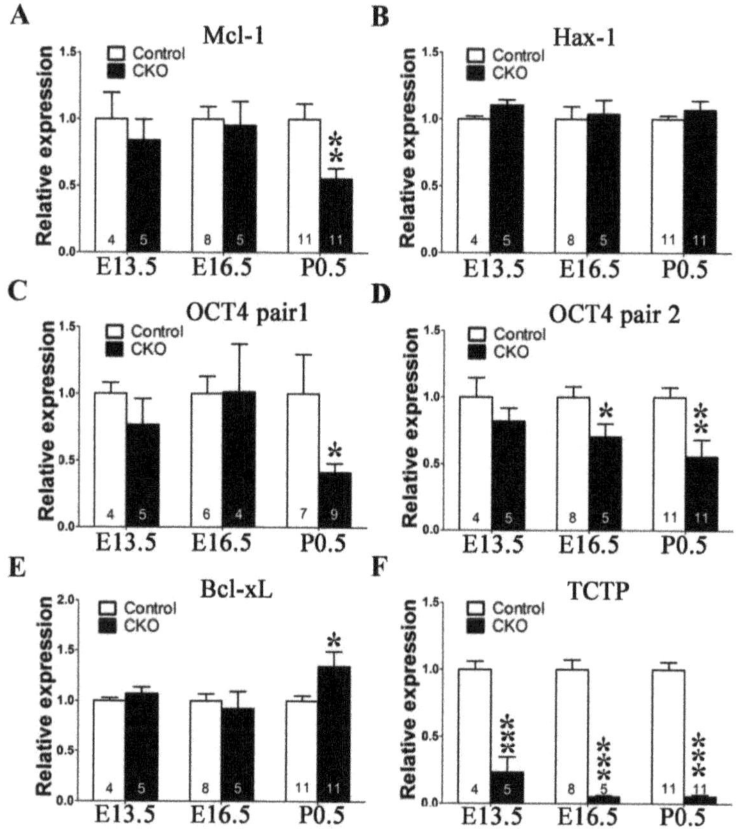

Figure 10. mRNA expression pattern of cell survival-related genes of TCTP-cKO mice compared with littermate controls. (**A–F**) Mcl-1, Hax-1, Oct-4 pair1, Oct-4 pair2, Bcl-xL, and TCTP mRNA expression from control and TCTP-cKO mice at E13.5, E16.5, and P0.5 were detected by qPCR. (Data presented as mean ± SEM, * $p < 0.05$, ** $p < 0.01$, *** $p < 0.001$, Student's t test, n = 4–11 per group).

Table 2. Primer sequences for qPCR gene expression analysis.

	Sequence	
Gene	**5′-Sense-3′**	**5′-Anti-Sense-3′**
Gapdh	CGACTTCAACAGCAACTCCCACTCTTCC	TGGGTGGTCCAGGGTTTCTTACTCCTT
Mcl-1	GGAAGTCCTCGCCTGCGTCA	AAACATGGTCGGACGCCGCA
Bcl-xL	AGGCAGGCGATGAGTTTGAA	CGGCTCTCGGCTGCTGCATT
Hax-1	GACCTTGCCTTCCCACTCTCCTGA	GTCCCTGCGACCCCCAATCTG
OCT4 pair1 *	GTGAGCCGTCTTTCCACCAGG	GGGTGAGAAGGCGAAGTCTG
OCT4 pair2 #	CCCTCCCTGGGGATGCTGTGAG	GAGTGACAGACAGGCCAGGCTCC
Nanog	TGCGGCTCACTTCCTTCTGACTTC	GGCCCTTGTCAGCCTCAGGAC
TCTP	TATATGAGGTTGGGGAGCGCCCG	CCTCCGGACCTTCAGCGGAA

* The pair of primers are located on exon 1 of OCT4 gene. # The pair of primers are located on exon 4 of OCT4 gene.

Taken together, our results demonstrate that TCTP is essential for cell survival during early neuronal and glial differentiation during CNS development. Thus, enhanced neuronal and functional loss and decreased cell numbers of Tuj1 and doublecortin-positive neurons mediated through Mcl-1, Bcl-xL, Oct4, and cyclin D2 and E2 suppression and caspase-3 cleavage activation resulting in increased apoptosis and decreased proliferation may contribute to the perinatal death of *TCTP* mutant mice.

4. Discussion

In this study, we generated and characterized mutant mouse strains with selective deletion of TCTP in neural precursor cells mediated by *Nestin*-driven Cre expression. We found (1) a dramatic phenotype of growth retardation, (2) early postnatal death, (3) enhanced neuronal loss and functional defect on Tuj1 and doublecortin-positive neurons, (4) increased apoptosis and decreased proliferation, and (5) decreased expression of mMcl-1, Bcl-xL, hax-1, and Oct4 in *Nestin-cre*-driven TCTP conditional knockout mice. Furthermore, we also demonstrated an obligatory role of TCTP in the maturation process of neuronal progenitor cell development.

In prenatal development, neurogenesis is responsible for populating the growing brain. New neurons are continually made throughout adulthood in predominantly two regions of the brain, the subventricular zone (SVZ) lining the lateral ventricles, where new cells migrate to the olfactory bulb via the rostral migratory stream, and the subgranular zone (SGZ), part of the dentate gyrus of the hippocampus. We observed high levels of TCTP protein expression in those regions in wild-type mice at P0.5. Moreover, loss of TCTP protein in the brain caused decreased BrdU-positive cells in neurogenetic regions, and the size of the lateral ventricle of the mutant telencephalon was significantly larger relative to littermate controls at P1. These results suggest that TCTP regulates prenatal brain development and is involved in new neuron proliferation. Importantly, TCTP has been shown to bind to Mcl-1 and Bcl-xL, antiapoptotic members of the Bcl-2 family [9,36,37]. TCTP is therefore closely associated with apoptotic processes. The TUNEL-positive apoptotic cell-prone nature of the TCTP-deficient brain was first noted at E13.5 and significantly increased at E16.5, with maximal apoptotic cells found at P0.5. We showed that loss of TCTP expression dramatically decreased Mcl-1 and Bcl-xL protein expression then increased cleaved caspase-3 protein levels in apoptotic cells and during neurogenesis on postnatal day 0.5. This phenotype, together with the proliferation defect observed in TCTP mutants, is sufficient to explain why the TCTP-null mice died early during neurogenesis. These results are consistent with the finding from the study of *Xenopus laevis* [21], which demonstrated that TCTP regulates retinal axon growth through the interaction with Mcl-1 and Bcl-xL protein to inhibit caspase-3 activation. Our study demonstrated in a mammalian model system that TCTP is also important for whole brain development and that this function is essential for the perinatal survival of mice. We also found that loss of TCTP in the brain caused a significant decrease in Hax-1 protein expression. Previous studies indicated that Hax-1 was involved in the regulation of apoptosis or programmed cell death [38]. Hax-1 and TCTP have been reported to regulate calcium homeostasis. Homozygous deletion of Hax-1 in mice results in excessive apoptosis of neurons and postnatal death caused by a loss of motor coordination and function, leading to a failure to eat or drink [38]. These phenotypes mimic TCTP mutant mice, suggesting that TCTP might interact with Hax-1 directly or indirectly.

On the other hand, Susini et al. proposed that TCTP could anchor the mitochondria and bind to MCL-1 and Bcl-xL to inhibit the dimerization of Bax and protect from apoptotic cell death by antagonizing the Bax function [33]. However, the antiapoptotic function of TCTP occurs in the mitochondria, where it inhibits Bax-induced damage. Our data did not show a significant difference in the Bax protein level with whole-cell lysates when TCTP was deleted. An analysis with a purified fraction of mitochondria in the brain tissue is needed to substantiate our results and clarify this issue. There is Mcl-1 pro-survival large isoform (Mcl-1L) and pro-apoptotic short isoform (Mcl-1S) in cells. We detected Mcl-1L but not Mcl-1S. Therefore, a more suitable antibody and well-prepared cell lysates are needed to clarify the changes of Mcl-1L and Mcl-1S in cKO mice. In addition, p53 may contribute to the cell death induced by *TCTP* deletion. Recent evidence has demonstrated that TCTP can facilitate Mdm2-mediated ubiquitination of P53 by competing with and preventing the binding of Numb, an inhibitor of Mdm2, to the Mdm2-P53 complex [39]. Further study is needed to clarify the role of P53 in the defect CNS development of *TCTP* mutant mice. Our study exploring the role of p53 in neurodevelopmental defects of *TCTP* mutant mice with double knockout mice ($Nestin^{Cre/+}$; $TCTP^{flox/flox}$; $p53^{-/-}$) is in progress. Our unpublished data showed that *P53* deficiency caused a more severe phenotype of $GFAP^{Cre/+}$; $TCTP^{flox/flox}$ knockout mice with $GFAP^{Cre/+}$; $TCTP^{flox/flox}$; $p53^{-/-}$ double knockout mice. In addition, Hsu et al. demonstrated that dTCTP functions upstream of mTOR-dS6K and regulates fly cell growth by positively regulating dRheb activity [40]. Our previous publication showed that the mTOR-dS6K pathway contributed to the regulation of TCTP in beta cells (17) but not fibroblasts (15) in a cell-type-specific manner.

In this study, we showed that TCTP is required for the proliferation, differentiation, and migration of neuronal progenitor cells. The *Nestin-cre*-derived TCTP disruption specific to neuronal progenitor cells caused the loss of Tuj1- or DCX-negative cells (Figure 5) and cell death in TCTP-negative cells (Figure 8) mediated through a cell autonomous mechanism. In contrast, it has been shown that TCTP could be secreted through the TSAP6 protein, which is also involved in exosome production [8]. Secreted TCTP also promotes liver regeneration and enhances colorectal cancer invasion, suggesting that TCTP protein functions as a cell nonautonomous regulator of cancer cell growth and proliferation [41,42]. Therefore, we could not exclude that the loss of TCTP in neuronal progenitor cells induces cell death through cell autonomous and nonautonomous mechanisms simultaneously.

A previous study showed that decreased levels of Oct-4 resulted in a failure to form the inner cell mass and cell number, lost pluripotency, and differentiated into troph-ectoderm [43,44]. Therefore, the level of Oct-4 expression in mice is vital for regulating pluripotency and early cell differentiation, including the brain, spinal cord, and nervous system, since one of its main functions is to keep embryonic stem cells from differentiating [44]. Oct-4 is also involved in the self-renewal of undifferentiated neuronal stem cells by forming a heterodimer with Sox2 followed by binding to DNA [45]. Low Oct-4 expression sustains self-renewal but is deficient in differentiation [46].

Our in vitro results showed that a deficiency of TCTP, specifically in neuronal progenitor cells, led to decreased survival of neuron progenitor cells and reduced Oct-4 protein expression in vivo, suggesting that disruption of TCTP reduces the capability of Oct-4 to regulate the self-renewal of neuron progenitor cells. Our results suggest that TCTP deficiency in neuronal progenitor cells leads to defects in transcription factor signaling, particularly impaired Oct-4 activation. This result is consistent with the finding that TCTP activates transcription of Oct-4 in *Xenopus laevis* oocytes [35] and kidney-derived stem cells [47] but not in mouse embryonic carcinoma P19 cells and J1 embryonic stem cells in vitro [48]. These results imply that Oct-4 may co-interact with TCTP protein and facilitate neuron cell differentiation. Furthermore, TCTP may mediate Oct-4-stimulated dependent or independent signals in neural stem cells (NSCs). The deficiency of TCTP causes the inhibition of NSC self-renewal with unidentified signaling pathways.

On the other hand, our previous study indicated that homozygous deletion of TCTP reduced cyclins D2 and E2 in the cell cycle G1 and S phases in developing embryos [15]. This result is consistent with selective deletion of TCTP in the brain, suggesting that TCTP plays a role in embryonic and brain cell proliferation via a mechanism that is linked to the regulation of the cell cycle machinery. More experiments will be required to address the molecular mechanisms involved in this interesting finding.

There was little or no milk in the stomachs of TCTP-deficient mice a few hours after birth. Previous studies indicated that the inability to feed resulted in neonatal death caused by the absence of nourishment but also because the liquid derived from milk is essential for homeostatic processes in newborns [49]. Our data show that TCTP deficiency triggering neonatal death was not rescued by oral administration of saline containing 15% glucose (Supplementary Table S1). Our data show that there was no difference in the cardiac function detected by an echocardiogram with a VisualSonics Vevo 660TM high-resolution imaging system between TCTP-deficient mice (*Nestin-cre*; $TCTP^{f/f}$) and littermate controls at E16.5 (unpublished results). Thus, the phenotype of early perinatal lethality of TCTP mutant mice was not caused by the heart problem. We also could not exclude the possibility that respiration failure may exist from the TCTP deficiency-induced defect in the respiration center in the pons and medulla. Several reports indicated that analysis of several neonatal lethal mutant mice revealed that nonfeeding or normal fasting newborns died 12 to 24 h after birth [49–51]. The TCTP mutant mice died between 24 and 36 h, suggesting that the inability to suckle milk was one possible cause of neonatal death. Feeding problems may be due to rejection by the mother. The major cause of the inability to feed in TCTP mutant mice might be associated with neuromuscular dysfunction.

Loss of TCTP in the developing brain might contribute to interference with normal physiological functions, such as the ability to suckle milk, olfactory sensitivity, movement coordination, and muscular function. Suckling is a complex process that consists of many structures of the brain and nerves, as well as the muscles required to extract milk [52]. According to our observation, TCTP-deficient mice can find and attach to a female mouse nipple, but we are not convinced whether these mice have a suckling response. The suckling response includes nipple attachment, suckling with rhythmic movement of the jaw and tongue, and the stretch response [53]. The interactions between motor and sensory neuronal pathways are linked to the central nervous system through the brain stem trigeminal complex. The data from Supplementary Figure S3 do not show a significant difference in the brain stem trigeminal ganglion between control and cKO mice detected by hematoxylin and eosin staining. To further confirm that the neurodevelopmental defects and phenotype of TCTP-cKO mice were due to the TCTP deficiency, $Nestin^{Cre/+}$; $TCTP^{f/+}$ mice will be crossed with hTCTP transgenic mice to produce double transgenic mice ($Nestin^{Cre/+}$; $TCTP^{f/+}$; *R26R-hTCTP*), which do not express the mouse TCTP protein but overexpress human TCTP, specifically in neuron progenitor cells. The phenotypes of the double transgenic mice will be compared with TCTP-cKO mice and *R26R-hTCTP* transgenic mice to check the rescue efficiency of hTCTP overexpression. Cortical neurons [25] and cortical progenitor cells [26] will be cultured for further study to explore whether disruption of TCTP sensitizes the neurons to apoptosis induced by DNA damage. On the other hand, Mcl-1 is a key regulator of apoptosis during CNS development [18]. *Nestin-cre*-derived knockout of TCTP decreased the expression of mMcl-1 (Figure 8C). Human Mcl-1 overexpression might rescue the early neonatal death of TCTP-cKO mutant mice.

Our results show new findings on the requirement for TCTP in the development and maintenance of neurons within the CNS. In vivo and in vitro results demonstrated that TCTP is required for neurogenesis and survival of the neuron maturation process. Additionally, TCTP regulates apoptotic cell death. Most importantly, our results implicate TCTP as a key regulatory molecule, involved in expanding the neural precursor pool and maintaining neuronal survival in the newborn mouse brain.

5. Conclusions

In our mouse model, TCTP deficiency in neuronal progenitor cells resulted in impaired neuronal and glial cell survival and differentiation through increased apoptosis and decreased proliferation, which led to perinatal death. We propose that TCTP is essential for cell proliferation, survival, and differentiation during CNS development. TCTP may be a target for the research of whole brain development and perinatal survival through the protection of neural precursor cells and neurons from apoptosis.

Supplementary Materials: The following are available online at http://www.mdpi.com/2073-4409/9/1/133/s1.

Author Contributions: Conceptualization, S.-H.C.; Methodology, C.-H.L., M.-J.T. and S.-H.C.; Validation, C.-H.L., M.-J.T. and S.-H.C.; Formal Analysis, C.-H.L., and M.-J.T.; Investigation, C.-H.L., M.-J.T. and S.-H.C.; Resources, S.-H.C.; Data Curation, C.-H.L., M.-J.T. and S.-H.C.; Writing—Original Draft Preparation, C.-H.L. and S.-H.C.; Writing—Review and Editing, S.-H.C.; Visualization, C.-H.L., M.-J.T. and S.-H.C.; Supervision, S.-H.C.; Project Administration, S.-H.C. Funding: S.-H.C. All authors have read and agreed to the published version of the manuscript.

Funding: This research was funded by National Science Council of Taiwan, NSC97-2314-B-320-009, NSC 98-2320-B-320-002-MY3, Ministry of Science and Technology, Taiwan, Grant MOST 105-2320-B-320-015, and by Tzu Chi University, Taiwan, Grant TCIRP99001-03Y1.

Acknowledgments: We thank the Transgenic Core Facility of Institute of Molecular Biology, Academia Sinica for generation of mutant mice with a conditional (floxed) allele of the TCTP gene ($TCTP^{f/+}$). We also thank the assistance of Ted H. Chiu (Department of Pharmacology, Tzu Chi University, Hualien, Taiwan) for editing the English of this manuscript.

Conflicts of Interest: The authors declare no conflict of interest.

References

1. Fiucci, G.; Lespagnol, A.; Stumptner-Cuvelette, P.; Beaucourt, S.; Duflaut, D.; Susini, L.; Amson, R.; Telerman, A. Genomic organization and expression of mouse Tpt1 gene. *Genomics* **2003**, *81*, 570–578. [CrossRef]
2. Bommer, U.A.; Thiele, B.J. The translationally controlled tumour protein (TCTP). *Int. J. Biochem. Cell Boil.* **2004**, *36*, 379–385. [CrossRef]
3. Gachet, Y.; Tournier, S.; Lee, M.; Lazaris-Karatzas, A.; Poulton, T.; Bommer, U.A. The growth-related, translationally controlled protein P23 has properties of a tubulin binding protein and associates transiently with microtubules during the cell cycle. *J. Cell Sci.* **1999**, *112*, 1257–1271.
4. Jung, J.; Kim, M.; Kim, M.J.; Kim, J.; Moon, J.; Lim, J.S.; Kim, M.; Lee, K. Translationally controlled tumor protein interacts with the third cytoplasmic domain of Na,K-ATPase alpha subunit and inhibits the pump activity in HeLa cells. *J. Boil. Chem.* **2004**, *279*, 49868–49875. [CrossRef] [PubMed]
5. Yarm, F.R. Plk phosphorylation regulates the microtubule-stabilizing protein TCTP. *Mol. Cell. Boil.* **2002**, *22*, 6209–6221. [CrossRef] [PubMed]
6. Cans, C.; Passer, B.J.; Shalak, V.; Nancy-Portebois, V.; Crible, V.; Amzallag, N.; Allanic, D.; Tufino, R.; Argentini, M.; Moras, D.; et al. Translationally controlled tumor protein acts as a guanine nucleotide dissociation inhibitor on the translation elongation factor eEF1A. *Proc. Natl. Acad. Sci. USA* **2003**, *100*, 13892–13897. [CrossRef] [PubMed]
7. Langdon, J.M.; Vonakis, B.M.; MacDonald, S.M. Identification of the interaction between the human recombinant histamine releasing factor/translationally controlled tumor protein and elongation factor-1 delta (also known as eElongation factor-1B beta). *Biochim. Biophys. Acta* **2004**, *1688*, 232–236. [CrossRef] [PubMed]
8. Amzallag, N.; Passer, B.J.; Allanic, D.; Segura, E.; Thery, C.; Goud, B.; Amson, R.; Telerman, A. TSAP6 facilitates the secretion of translationally controlled tumor protein/histamine-releasing factor via a nonclassical pathway. *J. Boil. Chem.* **2004**, *279*, 46104–46112. [CrossRef]
9. Liu, H.; Peng, H.W.; Cheng, Y.S.; Yuan, H.S.; Yang-Yen, H.F. Stabilization and enhancement of the antiapoptotic activity of mcl-1 by TCTP. *Mol. Cell. Boil.* **2005**, *25*, 3117–3126. [CrossRef]
10. Yang, Y.; Yang, F.; Xiong, Z.; Yan, Y.; Wang, X.; Nishino, M.; Mirkovic, D.; Nguyen, J.; Wang, H.; Yang, X.F. An N-terminal region of translationally controlled tumor protein is required for its antiapoptotic activity. *Oncogene* **2005**, *24*, 4778–4788. [CrossRef]
11. Rid, R.; Onder, K.; Trost, A.; Bauer, J.; Hintner, H.; Ritter, M.; Jakab, M.; Costa, I.; Reischl, W.; Richter, K.; et al. H2O2-dependent translocation of TCTP into the nucleus enables its interaction with VDR in human keratinocytes: TCTP as a further module in calcitriol signalling. *J. Steroid Biochem. Mol. Boil.* **2010**, *118*, 29–40. [CrossRef]
12. Rho, S.B.; Lee, J.H.; Park, M.S.; Byun, H.J.; Kang, S.; Seo, S.S.; Kim, J.Y.; Park, S.Y. Anti-apoptotic protein TCTP controls the stability of the tumor suppressor p53. *FEBS Lett.* **2011**, *585*, 29–35. [CrossRef]
13. MacDonald, S.M.; Rafnar, T.; Langdon, J.; Lichtenstein, L.M. Molecular identification of an IgE-dependent histamine-releasing factor. *Science* **1995**, *269*, 688–690. [CrossRef]
14. Kim, S.H.; Cairns, N.; Fountoulakisc, M.; Lubec, G. Decreased brain histamine-releasing factor protein in patients with Down syndrome and Alzheimer's disease. *Neurosci. Lett.* **2001**, *300*, 41–44. [CrossRef]
15. Chen, S.H.; Wu, P.S.; Chou, C.H.; Yan, Y.T.; Liu, H.; Weng, S.Y.; Yang-Yen, H.F. A knockout mouse approach reveals that TCTP functions as an essential factor for cell proliferation and survival in a tissue- or cell type-specific manner. *Mol. Boil. Cell* **2007**, *18*, 2525–2532. [CrossRef] [PubMed]
16. Wu, P.S.; Yang, C.Y.; Yen, J.J.; Chou, C.H.; Chen, S.H.; Wang, C.K.; Lai, Y.G.; Liao, N.S.; Yang-Yen, H.F. Critical roles of translationally controlled tumor protein in the homeostasis and TCR-mediated proliferation of peripheral T cells. *J. Immunol.* **2009**, *183*, 2373–2381. [CrossRef] [PubMed]
17. Tsai, M.J.; Yang-Yen, H.F.; Chiang, M.K.; Wang, M.J.; Wu, S.S.; Chen, S.H. TCTP is essential for beta-cell proliferation and mass expansion during development and beta-cell adaptation in response to insulin resistance. *Endocrinology* **2014**, *155*, 392–404. [CrossRef] [PubMed]
18. Arbour, N.; Vanderluit, J.L.; Le Grand, J.N.; Jahani-Asl, A.; Ruzhynsky, V.A.; Cheung, E.C.; Kelly, M.A.; MacKenzie, A.E.; Park, D.S.; Opferman, J.T.; et al. Mcl-1 is a key regulator of apoptosis during CNS development and after DNA damage. *J. Neurosci. Off. J. Soc. Neurosci.* **2008**, *28*, 6068–6078. [CrossRef]

19. Motoyama, N.; Wang, F.; Roth, K.A.; Sawa, H.; Nakayama, K.; Negishi, I.; Senju, S.; Zhang, Q.; Fujii, S.; Loh, D.Y. Massive cell death of immature hematopoietic cells and neurons in Bcl-x-deficient mice. *Science* **1995**, *267*, 1506–1510. [CrossRef]
20. Savitt, J.M.; Jang, S.S.; Mu, W.; Dawson, V.L.; Dawson, T.M. Bcl-x is required for proper development of the mouse substantia nigra. *J. Neurosci. Off. J. Soc. Neurosci.* **2005**, *25*, 6721–6728. [CrossRef]
21. Roque, C.G.; Wong, H.H.; Lin, J.Q.; Holt, C.E. Tumor protein Tctp regulates axon development in the embryonic visual system. *Development (Camb. Engl.)* **2016**, *143*, 1134–1148. [CrossRef]
22. Berube, N.G.; Mangelsdorf, M.; Jagla, M.; Vanderluit, J.; Garrick, D.; Gibbons, R.J.; Higgs, D.R.; Slack, R.S.; Picketts, D.J. The chromatin-remodeling protein ATRX is critical for neuronal survival during corticogenesis. *J. Clin. Investig.* **2005**, *115*, 258–267. [CrossRef] [PubMed]
23. Tronche, F.; Kellendonk, C.; Kretz, O.; Gass, P.; Anlag, K.; Orban, P.C.; Bock, R.; Klein, R.; Schutz, G. Disruption of the glucocorticoid receptor gene in the nervous system results in reduced anxiety. *Nat. Genet.* **1999**, *23*, 99–103. [CrossRef] [PubMed]
24. Magaud, J.P.; Sargent, I.; Clarke, P.J.; Ffrench, M.; Rimokh, R.; Mason, D.Y. Double immunocytochemical labeling of cell and tissue samples with monoclonal anti-bromodeoxyuridine. *J. Histochem. Cytochem. Off. J. Histochem. Soc.* **1989**, *37*, 1517–1527. [CrossRef] [PubMed]
25. Yen, C.L.; Mar, M.H.; Meeker, R.B.; Fernandes, A.; Zeisel, S.H. Choline deficiency induces apoptosis in primary cultures of fetal neurons. *FASEB J. Off. Publ. Fed. Am. Soc. Exp. Biol.* **2001**, *15*, 1704–1710. [CrossRef] [PubMed]
26. Callaghan, D.A.; Dong, L.; Callaghan, S.M.; Hou, Y.X.; Dagnino, L.; Slack, R.S. Neural precursor cells differentiating in the absence of Rb exhibit delayed terminal mitosis and deregulated E2F 1 and 3 activity. *Dev. Biol.* **1999**, *207*, 257–270. [CrossRef] [PubMed]
27. Gilyarov, A.V. Nestin in central nervous system cells. *Neurosci. Behave. Physiol.* **2008**, *38*, 165–169. [CrossRef]
28. Und Halbach, O.V. Immunohistological markers for staging neurogenesis in adult hippocampus. *Cell Tissue Res.* **2007**, *329*, 409–420. [CrossRef]
29. Lendahl, U.; Zimmerman, L.B.; McKay, R.D. CNS stem cells express a new class of intermediate filament protein. *Cell* **1990**, *60*, 585–595. [CrossRef]
30. Francis, F.; Koulakoff, A.; Boucher, D.; Chafey, P.; Schaar, B.; Vinet, M.C.; Friocourt, G.; McDonnell, N.; Reiner, O.; Kahn, A.; et al. Doublecortin is a developmentally regulated, microtubule-associated protein expressed in migrating and differentiating neurons. *Neuron* **1999**, *23*, 247–256. [CrossRef]
31. Gleeson, J.G.; Lin, P.T.; Flanagan, L.A.; Walsh, C.A. Doublecortin is a microtubule-associated protein and is expressed widely by migrating neurons. *Neuron* **1999**, *23*, 257–271. [CrossRef]
32. Liang, H.; Hippenmeyer, S.; Ghashghaei, H.T. A Nestin-cre transgenic mouse is insufficient for recombination in early embryonic neural progenitors. *Biol. Open* **2012**, *1*, 1200–1203. [CrossRef] [PubMed]
33. Susini, L.; Besse, S.; Duflaut, D.; Lespagnol, A.; Beekman, C.; Fiucci, G.; Atkinson, A.R.; Busso, D.; Poussin, P.; Marine, J.C.; et al. TCTP protects from apoptotic cell death by antagonizing bax function. *Cell Death Differ.* **2008**, *15*, 1211–1220. [CrossRef]
34. Niwa, H.; Miyazaki, J.; Smith, A.G. Quantitative expression of Oct-3/4 defines differentiation, dedifferentiation or self-renewal of ES cells. *Nat. Genet.* **2000**, *24*, 372–376. [CrossRef]
35. Koziol, M.J.; Garrett, N.; Gurdon, J.B. Tpt1 activates transcription of oct4 and nanog in transplanted somatic nuclei. *Curr. Boil. CB* **2007**, *17*, 801–807. [CrossRef] [PubMed]
36. Li, F.; Zhang, D.; Fujise, K. Characterization of fortilin, a novel antiapoptotic protein. *J. Boil. Chem.* **2001**, *276*, 47542–47549. [CrossRef] [PubMed]
37. Yang, T.; Buchan, H.L.; Townsend, K.J.; Craig, R.W. MCL-1, a member of the BLC-2 family, is induced rapidly in response to signals for cell differentiation or death, but not to signals for cell proliferation. *J. Cell. Physiol.* **1996**, *166*, 523–536. [CrossRef]
38. Suzuki, Y.; Demoliere, C.; Kitamura, D.; Takeshita, H.; Deuschle, U.; Watanabe, T. HAX-1, a novel intracellular protein, localized on mitochondria, directly associates with HS1, a substrate of Src family tyrosine kinases. *J. Immunol. (Baltim. Md. 1950)* **1997**, *158*, 2736–2744.
39. Amson, R.; Pece, S.; Lespagnol, A.; Vyas, R.; Mazzarol, G.; Tosoni, D.; Colaluca, I.; Viale, G.; Rodrigues-Ferreira, S.; Wynendaele, J.; et al. Reciprocal repression between P53 and TCTP. *Nat. Med.* **2011**, *18*, 91–99. [CrossRef]

40. Hsu, Y.C.; Chern, J.J.; Cai, Y.; Liu, M.; Choi, K.W. Drosophila TCTP is essential for growth and proliferation through regulation of dRheb GTPase. *Nature* **2007**, *445*, 785–788. [CrossRef]
41. Hao, S.; Qin, Y.; Yin, S.; He, J.; He, D.; Wang, C. Serum translationally controlled tumor protein is involved in rat liver regeneration after hepatectomy. *Hepatol. Res.* **2016**, *46*, 1392–1401. [CrossRef] [PubMed]
42. Xiao, B.; Chen, D.; Luo, S.; Hao, W.; Jing, F.; Liu, T.; Wang, S.; Geng, Y.; Li, L.; Xu, W.; et al. Extracellular translationally controlled tumor protein promotes colorectal cancer invasion and metastasis through Cdc42/JNK/ MMP9 signaling. *Oncotarget* **2016**, *7*, 50057–50073. [CrossRef] [PubMed]
43. Nichols, J.; Zevnik, B.; Anastassiadis, K.; Niwa, H.; Klewe-Nebenius, D.; Chambers, I.; Scholer, H.; Smith, A. Formation of pluripotent stem cells in the mammalian embryo depends on the POU transcription factor Oct4. *Cell* **1998**, *95*, 379–391. [CrossRef]
44. Zaehres, H.; Lensch, M.W.; Daheron, L.; Stewart, S.A.; Itskovitz-Eldor, J.; Daley, G.Q. High-efficiency RNA interference in human embryonic stem cells. *Stem Cells (Dayt. Ohio)* **2005**, *23*, 299–305. [CrossRef]
45. Rodda, D.J.; Chew, J.L.; Lim, L.H.; Loh, Y.H.; Wang, B.; Ng, H.H.; Robson, P. Transcriptional regulation of nanog by OCT4 and SOX2. *J. Boil. Chem.* **2005**, *280*, 24731–24737. [CrossRef]
46. Radzisheuskaya, A.; Chia Gle, B.; dos Santos, R.L.; Theunissen, T.W.; Castro, L.F.; Nichols, J.; Silva, J.C. A defined Oct4 level governs cell state transitions of pluripotency entry and differentiation into all embryonic lineages. *Nat. Cell Biol.* **2013**, *15*, 579–590. [CrossRef]
47. Jing, Y.; He, L.L.; Mei, C.L. Translationally-controlled tumor protein activates the transcription of Oct-4 in kidney-derived stem cells. *Exp. Ther. Med.* **2017**, *13*, 280–284. [CrossRef]
48. Cheng, X.; Li, J.; Deng, J.; Li, Z.; Meng, S.; Wang, H. Translationally controlled tumor protein (TCTP) downregulates Oct4 expression in mouse pluripotent cells. *BMB Rep.* **2012**, *45*, 20–25. [CrossRef]
49. Mizushima, N.; Yamamoto, A.; Hatano, M.; Kobayashi, Y.; Kabeya, Y.; Suzuki, K.; Tokuhisa, T.; Ohsumi, Y.; Yoshimori, T. Dissection of autophagosome formation using Apg5-deficient mouse embryonic stem cells. *J. Cell Boil.* **2001**, *152*, 657–668. [CrossRef]
50. DeChiara, T.M.; Vejsada, R.; Poueymirou, W.T.; Acheson, A.; Suri, C.; Conover, J.C.; Friedman, B.; McClain, J.; Pan, L.; Stahl, N.; et al. Mice lacking the CNTF receptor, unlike mice lacking CNTF, exhibit profound motor neuron deficits at birth. *Cell* **1995**, *83*, 313–322. [CrossRef]
51. Segre, J.A.; Bauer, C.; Fuchs, E. Klf4 is a transcription factor required for establishing the barrier function of the skin. *Nat. Genet.* **1999**, *22*, 356–360. [CrossRef] [PubMed]
52. Blass, E.M.; Teicher, M.H. Suckling. *Science* **1980**, *210*, 15–22. [CrossRef] [PubMed]
53. Westneat, M.W.; Hall, W.G. Ontogeny of feeding motor patterns in infant rats: An electromyographic analysis of suckling and chewing. *Behav. Neurosci.* **1992**, *106*, 539–554. [CrossRef] [PubMed]

cells

MDPI

Article

DHA Affects Microtubule Dynamics Through Reduction of Phospho-TCTP Levels and Enhances the Antiproliferative Effect of T-DM1 in Trastuzumab-Resistant HER2-Positive Breast Cancer Cell Lines

Silvia D'Amico [1,†], Ewa Krystyna Krasnowska [1,†], Isabella Manni [2], Gabriele Toietta [3], Silvia Baldari [3], Giulia Piaggio [2], Marco Ranalli [4], Alessandra Gambacurta [4], Claudio Vernieri [5,6], Flavio Di Giacinto [7], Francesca Bernassola [4], Filippo de Braud [5,8,‡] and Maria Lucibello [1,*,‡]

1 National Research Council of Italy, Institute of Translational Pharmacology (IFT-CNR), 00133 Rome, Italy; silvia.damico@ift.cnr.it (S.D.); ewa.krasnowska@ift.cnr.it (E.K.K.)

2 UOSD SAFU, Department of Research, Diagnosis and Innovative Technologies, IRCCS-Regina Elena National Cancer Institute, 00144 Rome, Italy; isabella.manni@ifo.gov.it (I.M.); giulia.piaggio@ifo.gov.it (G.P.)

3 Tumor Immunology and Immunotherapy Unit, IRCCS-Regina Elena National Cancer Institute, 00144 Rome, Italy; gabriele.toietta@ifo.gov.it (G.T.); silvia.baldari@yahoo.it (S.B.)

4 Department of Experimental Medicine, University of Rome "Tor Vergata", 00133 Rome, Italy; ranallim@uniroma2.it (M.R.); gambacur@uniroma2.it (A.G.); bernasso@uniroma2.it (F.B.)

5 Medical Oncology Department, Fondazione IRCCS Istituto Nazionale dei Tumori, 20133 Milan, Italy; Claudio.Vernieri@istitutotumori.mi.it (C.V.); Filippo.DeBraud@istitutotumori.mi.it (F.d.B.)

6 IFOM, the FIRC Institute of Molecular Oncology, 20139 Milan, Italy

7 Department of Neuroscience, Università Cattolica del Sacro Cuore, 00168 Roma, Italy; digiacintoflavio@gmail.com

8 Oncology and Hemato-Oncology Department, University of Milan, 20122 Milan, Italy

* Correspondence: maria.lucibello@ift.cnr.it

† These authors contributed equally to this work.

‡ These authors contributed equally to this work.

Received: 10 January 2020; Accepted: 16 May 2020; Published: 19 May 2020

Abstract: Trastuzumab emtansine (T-DM1) is an anti-human epidermal growth factor receptor 2 (HER2) antibody-drug conjugated to the microtubule-targeting agent emtansine (DM1). T-DM1 is an effective agent in the treatment of patients with HER2-positive breast cancer whose disease has progressed on the first-line trastuzumab containing chemotherapy. However, both primary and acquired tumour resistance limit its efficacy. Increased levels of the phosphorylated form of Translationally Controlled Tumour Protein (phospho-TCTP) have been shown to be associated with a poor clinical response to trastuzumab therapy in HER2-positive breast cancer. Here we show that phospho-TCTP is essential for correct mitosis in human mammary epithelial cells. Reduction of phospho-TCTP levels by dihydroartemisinin (DHA) causes mitotic aberration and increases microtubule density in the trastuzumab-resistant breast cancer cells HCC1954 and HCC1569. Combinatorial studies show that T-DM1 when combined with DHA is more effective in killing breast cells compared to the effect induced by any single agent. In an orthotopic breast cancer xenograft model (HCC1954), the growth of the tumour cells resumes after having achieved a complete response to T-DM1 treatment. Conversely, DHA and T-DM1 treatment induces a severe and irreversible cytotoxic effect, even after treatment interruption, thus, improving the long-term efficacy of T-DM1. These results suggest that DHA increases the effect of T-DM1 as poison for microtubules and supports the clinical development of the combination of DHA and T-DM1 for the treatment of aggressive HER2-overexpressing breast cancer.

Keywords: phospho-TCTP; DHA; T-DM1; HER2-positive breast cancer

1. Introduction

HER2-positive breast cancer (HER2+ BC), which represents about 25–30% of breast cancers, is characterized by the overexpression of the human epidermal growth factor receptor 2 (HER2/neu), a tyrosine kinase receptor (RTK). Therapies with monoclonal antibodies of high affinity specific for the HER2 receptor, or combinations of multiple anti-HER2 antibodies, have led to a significant improvement in therapeutic response; despite this, many patients develop resistance to therapy [1–3].

Trastuzumab emtansine (T-DM1) is an anti-HER2 antibody-drug linked to the anti-mitotic agent emtansine (DM1). T-DM1 treatment provides significant clinical benefit in breast cancer patients previously treated with chemotherapy and HER2-directed therapy, and in patients with HER2-positive early breast cancer who had residual invasive disease after completion of neoadjuvant therapy [4–7]. However, some patients may develop disease progression [8,9]. Identification of novel combination therapies for T-DM1 represents a major challenge to improve treatment effectiveness and to delay or prevent acquired resistance to HER2 inhibition.

Recently, we found that high levels of the nuclear phosphorylated form of Translationally Controlled Tumour Protein (phospho-TCTP) in HER2+ BC is associated with adverse prognostic factors and with a poor clinical response to trastuzumab therapy, suggesting a possible application of phospho-TCTP as a new marker for breast cancer [10].

TCTP is a highly conserved protein and it has been implicated in different physiological processes, including cell proliferation, cell shape, and resistance to stress [11–15]. Gene knockout studies have revealed that TCTP-deficient mice and TCTP-deficient mutants of Drosophila die early during embryogenesis, suggesting its implication in cell proliferation [12,16]. TCTP is a critical target in cancer therapy [15,17–19]. Moreover, TCTP has been identified as a critical regulator of the tumour suppressor p53 [14,20]. TCTP has been also described as a positive regulator of epithelial-to-mesenchymal transition [21,22]. Numerous clinical data show that TCTP overexpression is associated with tumour progression and poor clinical outcome in many poorly differentiated tumours [20,23–26]. Interestingly, in human breast cancer, high-TCTP status associates with poorly differentiated aggressive G3-grade tumours, predicting poor prognosis [20]. Moreover, TCTP may mediate many biological functions through the interaction with proteins involved in relevant function in cancer biology. Among these are: i) polo-like kinase 1 (PLK1), a member of the polo-like family of serine/threonine kinases that plays a crucial role in cell-cycle regulation during mitosis [27]; ii) Y-box-binding protein 1 (YBX1), a transcription and translation regulator protein that increase cancer cell invasiveness and spreading [28]; iii) the myeloid cell leukemia-1 protein (Mcl-1), an anti-apoptotic Bcl-2 family member [29]. Despite numerous reports suggesting important functions of TCTP in the context of tumour biology, its precise role is less clear.

Our previously reported data show that TCTP is a direct substrate of PLK1 in mammary carcinoma cells [10]. These data are in line with earlier published evidence [30–32]. Notably, the residues of Ser46 and Ser64 of TCTP, that are specifically phosphorylated by PLK1 [30], are located in a highly conserved intrinsically disordered loop of the protein, which contains a highly conserved TCTP signature. The phosphorylation event of TCTP by PLK1 may be critical for the activity of the protein, as it has been demonstrated that phospho-TCTP detaches from the spindle at the metaphase-to-anaphase transition [33].

It has been reported that TCTP regulates spindle microtubule dynamics during mitosis and its overexpression may lead to microtubule stabilization and alterations in cell morphology [34]. Conversely, TCTP phosphorylation by PLK1 increases microtubule dynamics by decreasing the microtubule-stabilizing activity of TCTP [30,33]. Consistent with this data, it has been reported that TCTP phosphorylating activity is low throughout the cell cycle, but increases in mitosis [11].

All together, these data suggest that TCTP may be a crucial player in mitotic processes and a fine equilibrium between the phosphorylated and non-phosphorylated forms of TCTP is required for maintaining the dynamic state of microtubules.

TCTP is a target of dihydroartemisinin (DHA) [10,35–37]. DHA, the active metabolite of Artemisinin, was discovered as an anti-malarial agent by Dr. Youyou Tu, who was awarded the 2015 Nobel Prize in Physiology or Medicine [38]. Notably, it has been reported that microarray-based mRNA expression of human TCTP is correlated with sensitivity to artesunate (a derivative of artemisinin) in tumour cells, suggesting that human TCTP contributes to the response of tumour cells to the drug [36]. Today, phase 1 clinical trials are underway showing that DHA has good safety and tolerability profile after long-term administration in patients with breast cancer or carcinoma of the uterine cervix [39–41]. These data are compatible with those obtained in malarial patients who, on the contrary, followed pharmacological treatments for shorter times [42,43]. All together, these findings shed light on the potential use of DHA as an anti-tumour agent and support the growing interest in this drug compound [44].

We have previously shown in HER2 overexpressing breast cancer cell lines that DHA, by reducing the expression levels of the phosphorylated form of TCTP, enhances the response to treatment with drugs as doxorubicin, cisplatin and trastuzumab.

Here we have investigated, in depth, the response to DHA in combination with T-DM1 in breast cancer cells resistant to trastuzumab therapy. Our data show that the combination treatments caused growth inhibition through the induction of severe mitotic perturbations, which in turn led tumour cells into an unstable state no longer compatible with viability.

2. Materials and Methods

2.1. Chemicals

Dihydroartemisinin was from Selleckem (Munich, D) T-DM1 was provided by Genentech (South San Francisco, CA, USA). N-Acetyl-L- Cysteine (NAC) by Sigma-Aldrich, St. Louis, MO, USA.

2.2. Cell Culture and Treatments

All cell lines were from ATCC. HCC1569 (ER−, Pr−, Her2+), HCC1954 (ER−, Pr−, Her2+) and BT-474 cells (ER+, Pr+/−, /Her2+) were maintained in RPMI-1640 or Dulbecco's modified Eagle's medium (DMEM) supplemented with L-glutamine, antibiotics and 10% heat-inactivated foetal bovine serum (FBS) all from Corning (New York, NY, USA), according to ATCC indications. MCF10A cells were maintained in Dulbecco's modified Eagle's medium DMEM/F-12 medium containing 5% horse serum (Thermo Fisher Scientific, Waltham, MA, USA) hydrocortisone (0.5 µg/mL) (Sigma-Aldrich), insulin (10 µg/mL) (Sigma-Aldrich), Epidermal growth factor, EGF (20 ng/mL) (Sigma-Aldrich), cholera toxin (100 ng/mL) (Sigma-Aldrich), penicillin (100 units/mL) and streptomycin (100 µlg/mL) (Corning).

Cells (at 5000 cells/cm^2 or otherwise indicated) were cultured in a humidified incubator in an atmosphere of 5% CO_2 at 37 °C. Before any experiment, cells were detached by mild trypsinization, washed, plated in medium containing 10% FBS, and allowed to recover for 24 h. Cells were treated with DHA dissolved in DMSO (Sigma-Aldrich) or with T-DM1. Control media contained the same amount of DMSO-vehicle (<0.1%). All cell lines were tested for mycoplasma contamination regularly using MycoAlert Mycoplasma Detection Kit (Lonza, Basel, Switzerland) and were authenticated by STR sequencing (BMR Genomics, Padoa, Italy).

2.3. Antibodies

Primary antibodies were purchased from commercial sources.

Immunofluorescence staining: anti-TCTP (Abcam, Cambridge, UK; #ab133568); anti-Phospho-TCTP (Ser46) (Cell Signalling Technology, Leiden, NL #5251); anti-Ki-67 (Neo Markers, Fremont, CA, USA #RM-9106-S0); anti-FLAG monoclonal antibody (DYKDDDDK tag), clone M2 (anti-FLAG M2) (Sigma-Aldrich, #F1804); anti- αTubulin (Sigma-Aldrich#T9026).

Western Blot analysis: anti-histamine releasing factor (HRF)/TCTP (MBL International, Woburn, MA, USA, #JMO99-3); anti-Phospho-TCTP (Ser46) (Cell Signalling Technology, #5251); anti-Poly(ADP-ribose)polymerase (PARP) (Cell Signalling Technology #9542); anti-cleaved caspase 3 (Cell Signalling Technology #9661); anti-phospho-protein kinase B (PKB/AKT hereafter referred as AKT, Ser473) (Cell Signalling Technology #9271); anti-AKT (Cell Signalling Technology, #9272); anti phospho-AMP-activated protein kinase (AMPK) (Thr172) (Cell Signalling Technology #2531); anti-AMPK (Cell Signalling Technology #2532); anti-cyclin B1 (Santa Cruz Biotechnology, Dallas, TX, USA, sc-245); anti-FLAG M2 (Sigma-Aldrich, #F1804); anti-phospho-Histone H2AX (Sigma-Aldrich, 05-636-25UG); anti-phospho Histone H3 (Ser10) (Merck-Millipore, #06-570); anti-p44/42 mitogen-activated protein kinase (MAPK) (Erk1/2) (Cell Signalling Technology#4695); anti-phospho p44/42 MAPK (Erk1/2) (Thr202/Tyr204)(Cell Signalling Technology, #9106); anti-β-actin (Sigma-Aldrich, A1978); anti-gliceraldeide 3-fosfato deidrogenasi (GAPDH) (Santa Cruz Biotechnology, #sc25778).

Flow cytometry analysis: purified mouse anti-human c-erbB-2 (BD Biosciences, Franklin Lakes, NJ, USA, #554300).

2.4. Cell Viability Assay

The ATP content was determined luminometrically by the CellTiter-Glo Luminescent Cell Viability assay, following the instructions of the manufacturers (Promega, Madison, WI, USA). Luminescence was measured with an automatic microtiter plate reader, Victor 3 V, Wallac 1420, Multilabel Counter (Perkin Elmer, Waltham, MA, USA).

2.5. ROS Production Assay

Reactive oxygen species (ROS) production was determined luminometrically by the ROS-Glo H_2O_2 Assay, following the instructions of the manufacturers (Promega). Luminescence was measured with an automatic microtiter plate reader, Victor 3 V, Wallac 1420, Multilabel Counter (Perkin Elmer).

2.6. Western Blot Analysis

Cells were washed with ice-cold phosphate-buffered saline (PBS), lysed in buffer contained 50 mM TRIS-HCl, pH 7.5, 400 mM NaCl, 10% glycerol, 0.5% NP40, 1% Tryton X-100 1 mM EDTA, 1 mM EGTA 2 mM DTT. All reagents were from Sigma-Aldrich. Buffer were supplemented with a protease inhibitor cocktail (P1860-Sigma-Aldrich) and phosphatase inhibitors cocktails (Sigma-Aldrich P5726 and Sigma-Aldrich P0044). Aliquots (20–60 μg) from total cell lysate proteins were resolved on 8–15% SDS–PAGE gels and analysed by immunoblotting with the indicated antibodies followed by decoration with peroxidase-labelled anti-rabbit (Thermo Fisher Scientific) or anti-mouse immunoglobulin G (IgG) (Dako-Agilent, Santa Clara, CA, USA) respectively. Blots were developed with enhanced chemiluminescence (ECL) Westar Supernova (Cyanagen, Bologna, Italy), following the instructions of the manufacturers.

2.7. Colony Formation Assays

Cells were seeded in 6-well plates at low density (1000 cells/plate), and cultured both in the absence or presence of drugs as indicated, for 6 days, and then incubated for 14 to 21 days to enable colony formation, after which they were fixed with 4% formaldehyde in PBS and stained with 0.1% crystal violet (Sigma-Aldrich). Relative quantification of colony number was performed using ImageJ Software (U.S. National Institute of Health, Bethesda, MD, USA).

2.8. Flow Cytometry

For cell cycle studies, cells were trypsinized and fixed for 45 min in methanol/acetone 4:1. After centrifugation at 950 RPM for 10 min, cells were stained with a solution containing 100 μg/mL RNase A and 50 μg/mL propidium iodide (Sigma-Aldrich) overnight in the dark at 4 °C. For detection of HER2 surface expression, cells were trypsinized and fixed for 10 min in 1% paraformaldehyde (Sigma-Aldrich). After blocking with 2% bovine serum albumin (BSA) (Sigma-Aldrich) in PBS, cells were probed with purified mouse anti-human c-erbB-2 (BD Biosciences) overnight in the dark at 4 °C. Primary antibody detection was obtained by reaction with secondary antibody Alexa Fluor 488 conjugated with IgG (Thermo Fisher Scientific, # A28175). Flow cytometry analysis was carried out using Fluorescence-activated cell sorting (FACS) Calibur flow cytometer (BD Biosciences). The percentage of cells in each stage of the cell cycle was determined using ModFit software (BD Biosciences). The percentage of cells in Sub-G1 phase was determined using FlowJo X (Tree Star Inc., Ashland, OR, USA).

2.9. *Immunofluorescence Staining*

Cells were grown on coverslips, fixed in 4% paraformaldehyde (Sigma-Aldrich) solution for 10 min. Then, samples were permeabilized in 0.2% Triton X-100/PBS for 5 min. After blocking with 3% BSA (Sigma-Aldrich) in PBS, cells were probed with the indicated primary antibodies and with Alexa Fluor 488-conjugated anti-rabbit IgG (Thermo Fisher Scientific, #A11034) or Alexa Fluor 555-conjugated anti-rabbit IgG (Thermo Fisher Scientific, #A-21428). Nuclei were counterstained with Hoechst (Fluka Biochemika, Buchs, Switzerland). Fluorescently labelled samples were imaged using a confocal LEICA TCS SP5 microscope (Leica, Heidelberg, Germany) equipped with an argon/krypton laser. Confocal sections were acquired at 0.4 μm intervals.

Excitation/emission wavelengths were 346/460 nm for Hoechst, 488/520 nm for Alexa Fluor 488 and 555/580 nm for Alexa Fluor 555. Images are shown as three-dimensional (3D) maximal projections reconstructed from z-stacks unless otherwise indicated. Magnification: 63× zoom 5 (bar = 5 μm), 63× (bar = 25 μm) and 20× (bar = 100 μm).

2.10. *Quantification of TCTP Distribution*

To study the distribution of the protein within cells, we acquired images at high magnification 63× zoom 5 (bar = 5 μm) of tumour cell lines stained with the indicated primary antibody and Hoechst (blue) at two different spectral range. Excitation/emission wavelengths used were 346/460 nm (first channel) for chromosome detection and 555/580 nm for protein detection (second channel). At least 12 cells were analysed for each sample. Thus, we performed a semi-automatic segmentation of chromosomes and cells by applying ImageJ plugin (Trainable Weka segmentation plugin) to both channels and then excluding all the cells at the image border. From these two binary masks (one for the whole cell and another for the chromosomes), we calculated pixel by pixel the distance from chromosomes of each point of the cells, using Matlab (Mathworks, USA). Collecting all the values of the second channel intensity for each distance from the chromosome, we obtained a line profile, which shows the distribution of the protein as a function of the distance from the chromosome. Signal intensity in the second channel is directly related to the quantity of protein. These results are based on a comparison between the fluorescence intensity for treated and untreated cells. The experimental setup has been kept constant for all acquisitions for cells of the same cellular line. Fluorescently labelled samples have been imaged using a confocal LEICA TCS SP5 microscope (Leica, Heidelberg, Germany) equipped with an argon/krypton laser.

2.11. *Quantification of Ki-67 Positive Cells*

We quantified Ki-67 positive cells from the analysis of fluorescence images deriving from a sample stained with Hoechst (blue) for Nuclei (first channel) and Ki-67 (red) (second channel). Excitation/emission wavelengths used were 346/460 nm for first channel and 555/580 nm for second channel. The signals from these fluorophores, acquired in two different spectral channels, allowed to contextually count the whole population of cells and distinguish the KI-67 positive cells. The count and segmentation of the whole cell population from the signal in the first channel have been performed using a custom made program of image analysis (Matlab, Mathworks, USA) based on a Gaussian filter for the background remotion and a watershed algorithm for cell segmentation. The algorithm produced a binary mask of the cells in the original image. By applying this mask to the image of the second channel, we calculated a mean value of signal intensity for each cell, and then distinguished two different groups within the whole population, through the set of a threshold value in intensity. In each group, we analysed: i) 2380 cells for MCF10A-pBabe; ii) 3612 cells for MCF10A-AATCTP; iii) 2483 cells for MCF10A-WTTCTP (right panel).

2.12. *Quantification of Microtubule Density*

For the quantitative analysis of the microtubule density, we analysed the fluorescence images obtained by staining the sample with Hoechst (blue) and αTubulin (green). Excitation/emission wavelengths used were 346/460 nm (blue) and 488/520 nm (green). At least 500 cells were analysed for each sample. For quantification analysis, we firstly removed the background to highlight the strong signal from tubulin by applying a Gaussian filter to images, and then we subtracted this image from the original one. Afterwards, we applied a threshold to segment the tubulin and create a binary image of these structures. Finally, we compared this mask to the mask of the whole area in the image field, which is covered by cell. Thus, to obtain a quantification of tubulin density, we divided these two values, obtaining the fraction of pixels, which appears to be crossed by tubulin structures.

2.13. *Growth Curve*

Cells (5×10^4) were seeded in each well of 6-well plate. The cell number for each cell line was counted each day for a 6-day time course; this number has been normalized on cell number counted 24 h after seeding to allow cells to recover from the trypsinization. Cell numbers have been counted in triplicate in three independent experiments. Doubling times were determined by fitting the data points to the exponential growth function in GraphPad Prism 5.01 software.

2.14. *Vector Construction*

Plasmids pcDNA3-TCTP were constructed as previously described [10].

2.15. *Mutagenesis*

Mutagenesis of sites of phosphorylation of TCTP Ser46 and Ser64 were performed as previously described [10].

2.16. *Recombinant Retroviral Vectors*

The PCR amplified fragments wild type (WT) and Ser46Ala Ser64Ala double mutant (AA) were cloned in *EcoR1* site of pBABE-Puro retroviral vector to obtain FLAG-TCTP-pBABE and FLAG-AA-TCTP-pBABE. All constructs were confirmed by DNA sequence analysis.

2.17. *Cell Transfection*

Retroviruses were produced by transfection of Phoenix-Ampho packaging cells with pBABE-puro, AA-TCTP-pBABE, and WT-TCTP-pBABE using Lipofectamine 2000 (Invitrogen, Carlsbad, CA, USA). At 48 h after transfection, supernatants containing the retroviral particles were collected and frozen at −80 °C until use. MCF10A cells were infected with diluted supernatant in the presence of 8 μg/mL Polybrene (Sigma-Aldrich) overnight, and cells containing the pBABE, AA-TCTP-pBABE, and WT-TCTPpBABE constructs were selected with puromycin (1 μg/mL) (Sigma-Aldrich) 48 h after infection. After 10 days in selective medium, the three pools referred to empty vector (MCF10A-pBABE), the wild type TCTP protein (WT-TCTP), the Ser46Ala Ser64Ala double mutant TCTP (AA-TCTP), were isolated. The puromycin selective pressure was removed 24 h before experimental procedures.

2.18. Evaluation of Cell Sensitivity to Combined Treatment

Cells were plated in triplicate in 96-well and treated with DHA, T-DM1, and with the DHA/T-DM1 combination. Growth inhibition was calculated as the percentage of viable cells compared to untreated cells by the CellTiter-Glo Luminescent Cell Viability assay (Promega, Madison, WI, USA) The CompuSyn software program has been used to calculated synergistic, additive or antagonistic effects. This program is based on the Median-Effect Principle (Chou) and the Combination Index–Isobologram Theorem (Chou-Talalay) [45]. Because all terms in the equations are ratios, all the dose units become dimensionless quantities. Drug can be different units. The combination index (CI) indicates a quantitative measure of the degree of drug interaction in terms of synergistic (CI < 1), additive (CI = 1) or antagonistic effect (CI > 1). DRI is the dose-reduction index and it is a measure of how many-fold the dose of each drug in a synergistic combination may be reduced at a given effect level compared with the doses of each drug alone.

2.19. Immunodeficient Mice Study

We generated HCC1954 cells expressing luciferase in order to implement bioluminescent imaging analysis to follow breast tumour growth in small animal models in vivo. Briefly, HCC1954 cells were transduced at multiplicity of infection MOI 10 with a third-generation self-inactivating lentiviral vector expressing firefly luciferase [46]. Six-week-old CB17SCID female mice were purchased from Charles River (Calco, Italy) and housed with laboratory chow and water available ad libitum.

A cell-line derived orthotopic xenograft model of breast cancer was established by mammary gland implantation of 5×10^5 HCC1954 luciferase-expressing cells. Mice were regularly palpated and tumour dimensions were measured once a week using a digital calliper. Moreover, tumour cell engraftment and early detection of tumour growth was assessed by longitudinal bioluminescent analysis (BLI). BLI analysis has been performed using the IVIS® Lumina II equipped with the Living Image® software for data quantification (PerkinElmer). Animals were sedated and D-luciferin (PerkinElmer) dissolved in PBS (150 mg/kg body weight) was administered i.p. 10 min before analysis [47]. Photons emitted from luciferase expressing HCC1954 cells implanted into the animals were collected with final accumulation times ranging from of 1 s to 1 min, depending on the intensity of the bioluminescence emission. All animal experiments were conducted in accordance with institutional guidelines, in the full observation of the Directive 2010/63/UE.

2.20. Statistical Analysis

All experiments were done at least three times unless otherwise indicated. The results are presented as means ± SD. Results were analysed using a Mann–Whitney test. One-way ANOVA followed by the Bonferroni test using the PRISM GraphPad software was used in the analysis of three or more data sets. Differences were considered significant for $P < 0.05$ and highly significant for $P < 0.01$ and $P < 0.001$

3. Results

3.1. DHA Affects Mitosis of HER2+ BC Cell Lines with Aberrant PI3K/AKT Signalling

We investigated the effect of DHA on HER2+ breast cancer cells resistant to trastuzumab. Since PI3KCA mutations and/or loss of phosphatase and tensin homolog (PTEN) have been associated with a lower response to trastuzumab and chemotherapy [1,48,49], we chose the HCC1954 cell line characterized by a mutation in the catalytic domain (H1047R) of *PI3KCA* and the HCC1569 cell line, which is PTEN-null. Both cell lines showed resistance to trastuzumab or pertuzumab therapy, in line with the data in literature [50], whereas HER2-positive breast cancer BT-474 cell line was sensitive to antibody therapy (Supplementary Figure S1).

Cells were grown in the presence or absence of DHA. As shown in Figure 1 (panel a), DHA induced a mild but significant inhibition of cell viability during the first 24–48 h at concentrations ranging from

1.25 to 5 μM, which are achievable in the clinic [40,42,43,51]. Western blot analysis also showed that DHA at these concentrations induced a reduction of both total and phospho-TCTP levels, in accordance with our previous data (Figure 1b) [10]. PLK1 inhibition impairs TCTP phosphorylation [10,52] as does DHA. Since PLK1 is a master mitotic regulator [27], we investigated the impact of DHA on mitotic cells. To this end, we performed a morphological analysis in both cell lines after 24 h of exposure to DHA. Control mitotic cells were characterized by a bright peri-chromosomal localisation of both phospho-TCTP (Figure 1c, left panel), and TCTP (Figure 1c, right panel), in accordance with data from the literature [31]. The signal intensity of phospho-TCTP was higher around the chromosomes than in the rest of cells, only in untreated cells, as indicated by a quantification analysis of peri-chromosomal localization of phospho-TCTP in both DHA-treated and untreated cells (Figure 1d). No differences in the signal intensity of TCTP were observed between untreated and treated cells (data not shown).

Interestingly, DHA-treated cells showed a significant increase of aberrant spindle structures (Figure 1e), such as multipolar or monopolar spindles, or chromosome misalignments (Figure 1c). The treatment did not arrest the progression of these cells through G2 and mitosis (Figure 1f). However, the cell cycle distribution analysis showed that DHA blocked the G1/S transition of a small, yet remarkable, percentage of HCC1569 cells (Figure 1f, upper panel). In addition, Western blot analysis showed that DHA induced an increase of Cyclin B1 and phospho-Histone-H3 in HCC1954 cells, thus suggesting that a delay in mitotic progression could be induced by treatment (Figure 1b). Altogether, these data show that aberrant mitosis could be induced by an early treatment with DHA in aggressive HER2+ BC cell lines and a reduction of the peri-chromosomal localization of phospho-TCTP level occurs at the same time.

3.2. DHA Induces a Decrease in AKT Phosphorylation Levels and DNA Damage Through the Increase of ROS in HER2+ BC Cell Lines

Then we investigated whether DHA could prevent phosphorylation on AKT's activation sites, as it has been demonstrated that active AKT is a crucial player in the downstream HER2 signalling pathways [1]. To this end, we evaluated the level of active AKT, in both cell lines after 24 h of DHA treatment. Western blot analysis of the total cell lysates showed that DHA induced a decrease in AKT phosphorylation levels (Figure 2c). DHA contains an endoperoxide moiety [53] that generated reactive oxygen species (ROS) in the first hours of treatment (Figure 2, panel a). To determine the implication of oxidative stress in the cytotoxicity of DHA, cells were pretreated with the ROS scavenger N-Acetyl-L-Cysteine (NAC) for 3 h followed by treatment with DHA. DHA-induced increase in ROS levels was abrogated by NAC (Figure 2a). In addition, NAC administration partially protected the cells from the anti-proliferative effect of DHA (Figure 2b), thus indicating that DHA cytotoxicity may be mediated at least in part by increase of ROS.

Then we investigated whether the phosphorylation of AKT can also be regulated by ROS. Notably, NAC reversed the inhibition of AKT phosphorylation induced by DHA treatment in both cell lines (Figure 2c). Since the increase of ROS levels may induce oxidative DNA damage in DHA-treated cells, we examined the phosphorylation levels of histone 2AX (γH2AX), as a marker of DNA double-strand breaks and genomic instability, in both cell lines after 24 h of exposure to DHA. A mild increase of H2AX phosphorylation at Ser139 was clearly detectable at 24 h of DHA treatment in both cell lines and it was reverted by NAC (Figure 2c).

These data show that, beyond the induction of mitotic perturbations, through the increase of oxidative stress, DHA could further damage cancer cells and, thus, might render them more sensitive to a subsequent treatment.

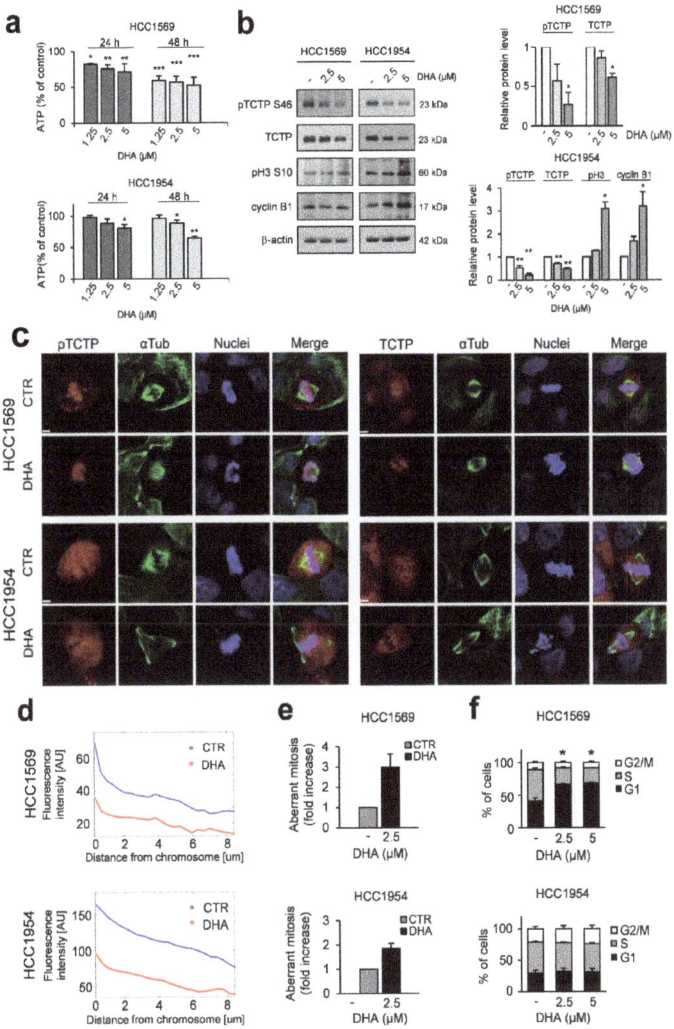

Figure 1. Dihydroartemisinin (DHA) induces aberrant mitosis in HER2-positive breast cancer (HER2+ BC) cell lines. (**a**) Cell viability was assessed by ATP assay. Data are expressed as the percentage of viable cells relative to controls. Values represent the mean ± SD, n = 3. (**b**) Western Blot analysis of the indicated proteins in cell lysates of cells treated with DHA 24 h. β-actin was used as loading control (right panel). For densitometric analysis, the intensity of each band was normalized to the respective β-actin (left panel). (**c**) Immunofluorescence detection of phospho-Translationally Controlled Tumour Protein (pTCTP S46) (red) or TCTP (red) and αTubulin (green) in cells treated with DHA at 2.5 μM for 24 h. Nuclei were stained with Hoechst (blue). The overlay of the three fluorochromes is shown (Merge), bar = 5 μm. Data from a representative experiment of three with similar results are shown. (**d**) Quantification of p-TCTP. Cells were treated as described in (**c**). The line profile shows the distribution of pTCTP as a function of the distance from the chromosome. The quantification was performed as described in Materials and Methods. (**e**) Bar graphs show relative quantification of fraction of cells with aberrant mitosis. At least 1000 cells were analysed in each group. Values represent the mean ± SD. (**f**) Cell cycle distribution in cells treated with DHA for 24 h. Values represent the mean ± SD, $n = 3$. * = $p < 0.05$, ** = $p < 0.01$, *** = $p < 0.001$ vs. control.

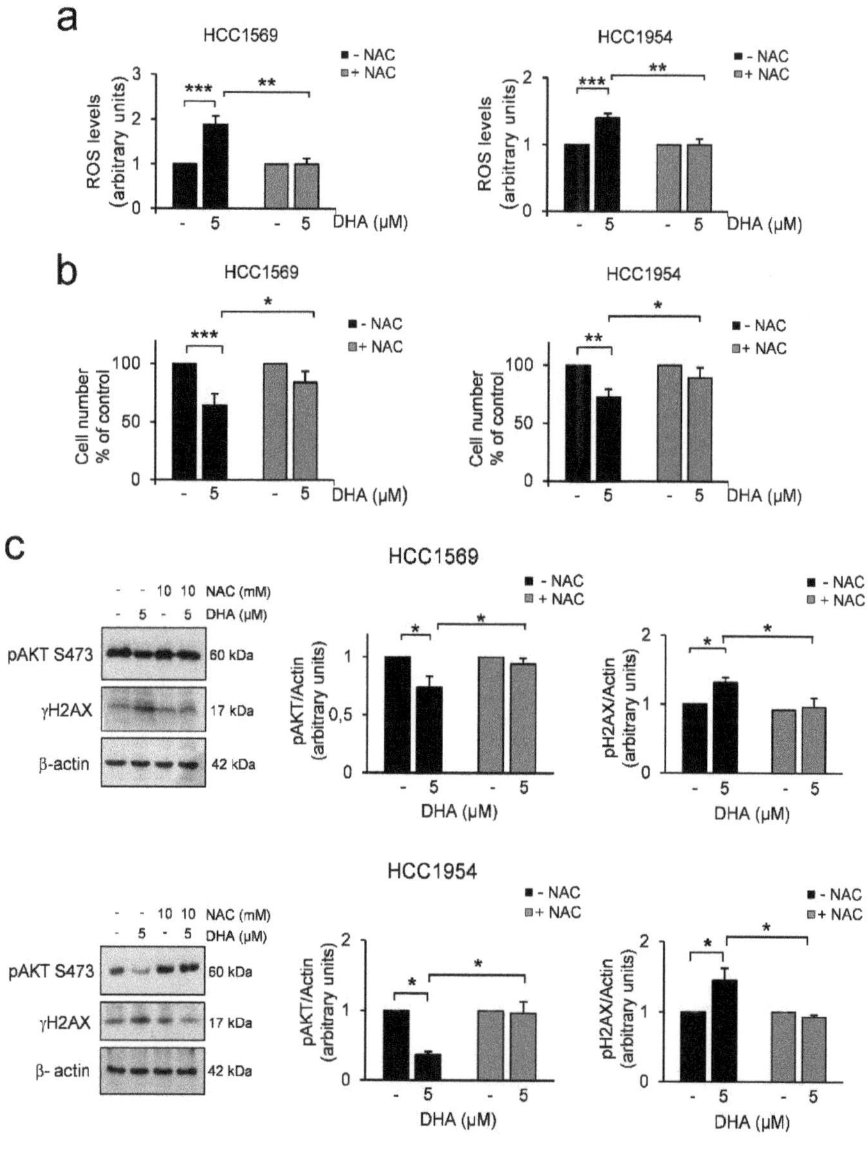

Figure 2. DHA induces oxidative stress in HER2+ BC cell lines. (**a**) Reactive oxygen species (ROS) production was assessed in cells pretreated with N-Acetyl-L- Cysteine (NAC) (10mM) followed by treatment with DHA for 24 h. ROS production was measured at the end of incubation time. Data are expressed as ROS levels relative to controls. n = 5. (**b**) Trypan blue exclusion assay was performed in cells treated as described in a). Data are expressed as the percentage of viable cells relative to controls. Values represent the mean ± SD, n = 5. (**c**) Western Blot analysis of the indicated proteins in HCC1569 (upper panel) and HCC1954 (lower panel) cells treated as described in a). β-actin was used as loading control. For densitometric analysis, the intensity of each band was normalized to the respective β-actin. * = $p < 0.05$, ** = $p < 0.01$, *** = $p < 0.01$.

3.3. Phosphorylation of TCTP is Required for Correct Mitotic Progression in Human Mammary Cells

To investigate the role of phospho-TCTP in cells committed to mitosis, we performed a series of experiments using the MCF10A cells, a non-tumorigenic human mammary cell line, which expresses low phospho-TCTP and TCTP levels (as previously demonstrated [10]), in which we overexpressed WT FLAG-tagged TCTP protein (WT) or a non-phosphorylatable mutant (AA) TCTP protein. To this aim, we carried out a retroviral infection to establish stable MCF10A subclones expressing: 1) empty vector, hereafter called MCF10A-pBabe; 2) wild-type TCTP protein (WT-TCTP), hereafter called MCF10A-WTTCTP; 3) Ser46Ala Ser64Ala double mutant TCTP (AA-TCTP), hereafter called MCF10A-AATCTP. As shown in Figure 3a, the expression of both forms (WT-TCTP and AA-TCTP) was verified assessing the level of tagged proteins by Western Blot (Figure 3a, left panel) and immunofluorescence analysis (Figure 3a, right panel).

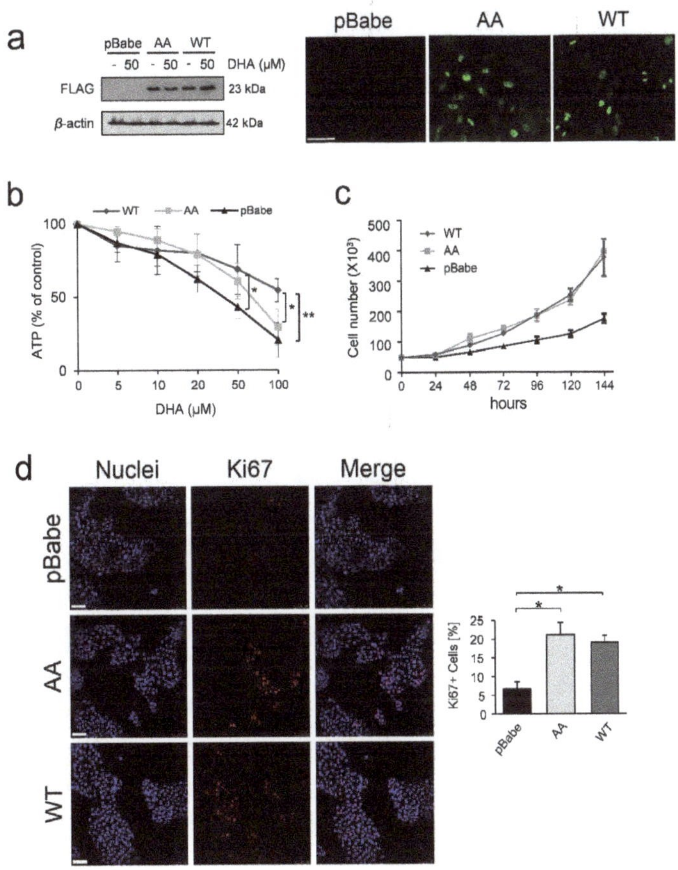

Figure 3. *Cont.*

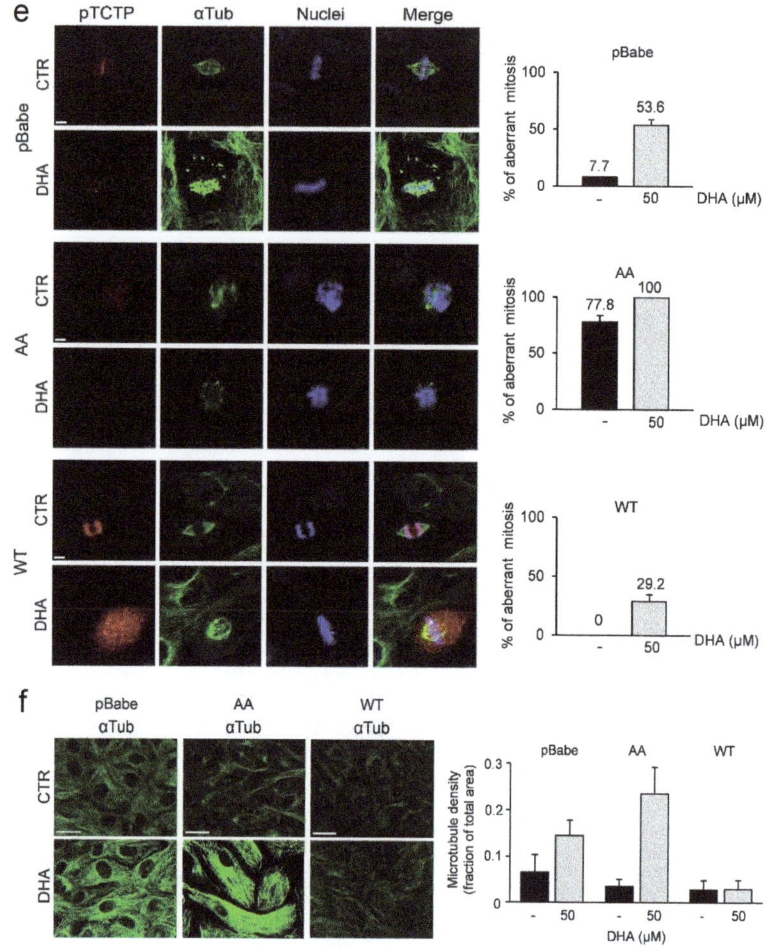

Figure 3. Effects of DHA on mitosis in human mammary cells. (**a**) Western blot analysis of FLAG-tagged TCTP protein in cell lysates of MCF10A-pBabe (pBabe), MCF10A-AATCTP (AA), MCF10A–WTTCTP (WT) cells after 6 days of exposure to DHA. β-actin was used as loading control (upper panel). Immunofluorescence detection of FLAG (green) fusion proteins, bar = 25 μm. (**b**) Cell viability was assessed by ATP assay in cells treated with DHA for 6 days. Data are expressed as the percentage of viable cells relative to controls. Values represent the mean ± SD, n = 3. * = $p < 0.05$, ** = $p < 0.01$. (**c**) Growth curve for all cell clones. (**d**) Immunofluorescence detection of Ki-67 (red). Nuclei were stained with Hoechst (blue). The overlay of the two fluorochromes is shown (Merge). Images are shown as single slice of a projection, bar = 100 μm. Data from a representative experiment of two with similar results are shown (left panel). Quantification of Ki-67 positive cells was performed as described in Materials and Methods (right panel). Values represent the mean ± SEM, * = $p < 0.05$. (**e**) Immunofluorescence detection of phospho-TCTP (pTCTP S46) (red) and αTubulin (green). Nuclei were stained with Hoechst (blue). The overlay of the three fluorochromes is shown (Merge). Cells were treated with DHA (50 μM) for 6 days. Bar = 5 μm). Data from a representative experiment of three with similar results are shown (left panel). Quantification of fraction of cells with aberrant mitosis. At least 1000 cells were analysed in each group. Values represent the mean ± SD. (right panel). (**f**) Immunofluorescence detection of αTubulin (green) in cells treated as described in (**e**), bar = 25 μm), left panel. Quantitative analysis of the microtubule density. The quantification was performed as described in Materials and Methods. Values represent the mean ± SD, right panel.

We compared MCF10A-pBabe cells to MCF10A-AATCTP and MCF10A-WTTCTP cells for their resistance to DHA. To this end, all cell clones were grown in the absence or presence of different concentrations of DHA for 6 days. As shown in Figure 3b, DHA induced an inhibitory effect in all three cell clones at significantly higher concentration than those active on cancer cells (>10 μM) (Table 1), suggesting that DHA can target cancer cells without harming healthy tissues. Overexpression of TCTP protected cells against DHA induced-cytotoxicity, as indicated by the fold increase in half-maximal effective concentration EC_{50} values reported in Table 1.

Table 1. Effects of DHA on cell growth in human mammary cells and in HER2+ BC cell lines. MCF10A-pBabe, MCF10A-AATCTP and MCF10A-WTTCTP cells were treated as described in legend 3b. HCC1569 and HCC1954 cells were treated as described in legend 4a. Half-maximal effective concentration (EC_{50}) values derived from concentration-response curves to DHA. Values represent the mean ± SD, $n = 3$. Significant differences between EC_{50} values from breast cancer cell lines and MCF10A cell clones are indicated, *** = $p < 0.01$.

Cell Line	EC_{50} (μM)
MCF10A-pBabe	31.65 ± 8.81
MCF10A-AATCTP	65.95 ± 11.08
MCF10A-WTTCTP	133.00 ± 30.45
HCC1569	8.50 ± 1.70 ***
HCC1954	7.61 ± 1.86 ***

We also found that overexpression of TCTP conferred a growth advantage. A growth curve analysis clearly demonstrate that doubling time of MCF10A subclones overexpressing both WTTCTP (46.79 h) and AATCTP (47.36 h) was shorter than that of MCF10A pBabe (71.88 h), indicating a more rapid proliferation (Figure 3, panel c). Moreover, quantification immunofluorescence analysis showed that both MCF10A-AATCTP and MCF10A-WTTCTP cells had higher expressions of Ki-67, a marker of cell proliferation, which is not detected in non-cycling cells (Figure 3d).

However, the overexpression of the non-phosphorylatable form of TCTP leads to an increase of a phenotype characterized by mitotic aberration [30]. In line with this data, control MCF10A-AATCTP cells showed aberrant mitotic figures, as shown by an immunofluorescence analysis (Figure 3e, left panel) and by a quantification analysis (Figure 3e, right panel). DHA at a concentration of 50 μM induced in these cells a severe damage, as indicated by a quantitative analysis showing an increase in microtubule mass mainly in MCF10A-AATCTP cells (Figure 3f). In contrast, we did not find any relevant alterations to the spindle morphology or to microtubule density between treated and non-treated cells in MCF10A-WTTCTP (Figure 3e,f) suggesting that expression of the phosphorylated form of TCTP is required for correct mitosis. DHA induced an increase in ROS levels in all cell clones. However, the increase in ROS levels after treatment was significantly higher in MCF10A-AATCTP cells when compared to MCF10A-WTTCTP cells (Supplementary Figure S2, panel a). The phosphorylation of AKT was not substantially affected by DHA and only a very light inhibition was observed in AA-MCF10 cells (Supplementary Figure S2, panel b). Moreover, DHA induced a very light increase of γH2AX levels only in MCF10A-AATCTP cells in comparison to the other cell clones (Supplementary Figure S2, panel b), thus indicating a greater vulnerability of these cells to DNA damage.

Altogether, these data show that TCTP is a crucial player in mitotic processes. By reduction of the phosphorylated form of TCTP, by mutagenesis and by DHA treatment, resulted in mitotic aberrations and increased microtubule density.

3.4. DHA Enhances T-DM1 Efficacy in Breast Cancer Cells Resistant to Trastuzumab Therapy

We then investigated the combinatorial effects of DHA and T-DM1 in both HCC1954 and HCC1569 cell lines. In a pilot study, we used two different protocols. In the first protocol, the effect of T-DM1 was studied in cells pre-treated with DHA and then exposed to T-DM1, while in the second protocol the drugs were administered at the same time. However, no remarkable differences in term of efficacy and CI values were obtained between the two protocols in both cell lines (data not shown). Since DHA induced mitotic aberration during the first 24 h of exposure, we decided to follow the first protocol.

HCC1954 cells were sensitive to both T-DM1 and DHA treatments as indicated by parameters from the dose-response curves obtained with various concentrations of these drugs (Table 2). Notably, T-DM1 showed a sharp increase in the slope of a dose-response curve, indicating a high cytotoxicity above the EC_{50} value.

Table 2. DHA in combination with T-DM1 causes synergistic inhibition of growth in HER2+ BC cancer cell lines. Cell viability was assessed by ATP assay in cells treated as described in legend 4a. Fractional inhibition = fraction decreased cell viability after treatment, control cells were set to "1". The parameters m, Dm, and r are the shape of the dose-effect curve, the potency (EC_{50}), and the conformity of the data to the mass-action law, respectively. CI values below 0.9 indicate synergistic effect. CI values = 1 indicate additive effect. DRI (DHA) and DRI (T-DM1) are the dose reduction index for DHA and T-DM1, respectively. Value represent the mean ± SD, n = 3.

HCC1954 cells						**HCC1569 cells**					
DHA (μM)		**Fractional inhibition**	**T-DM1 (μg/mL)**	**Fractional inhibition**		**DHA (μM)**		**Fractional inhibition**	**T-DM1 (μg/mL)**	**Fractional inhibition**	
(D_1)			(D_2)			(D_1)			(D_2)		
1		0.10 ± 0.08	0.01	0.04 ± 0.03		1.25		0.11 ± 0.06	1	0.03± 0.04	
2.5		0.15 ± 0.09	0.1	0.22 ± 0.07		2.5		0.26 ± 0.07	2.5	0.09 ± 0.07	
5		0.29 ± 0.14	0.25	0.38 ± 0.15		5		0.41 ± 0.06	5	0.13 ± 0.08	
10		0.61 ± 0.16	0.5	0.68 ± 0.13		10		0.54 ± 0.05	10	0.23 ± 0.02	
20		0.85 ± 0.08	1	0.78 ± 0.07		20		0.65 ± 0.03	20	0.21 ± 0.12	
									50	0.57 ± 0.11	
									200	0.59 ± 0.04	
Parameter			Parameter			Parameter			Parameter		
EC_{50} (μM)		7.61 ± 1.86	EC_{50} (μg/mL)	0.33 ± 0.18		EC_{50} (μM)		8.5 ± 1.7	EC_{50} (μg/mL)	101.70 ± 44.8	
m		1.88 ± 0.47	m	0.88 ± 0.26		m		0.90 ± 0.13	m	0.78 ± 0.24	
r		0.98 ± 0.01	r	0.96 ± 0.02		r		0.98 ± 0.02	r	0.97 ± 0.01	
		(D1) + (D2) Ratio 10:1						(D_1) + (D_2) Ratio 1:1			
		Fractional inhibition	CI	DRI (D_1)	DRI (D_2)			Fractional inhibition	CI	DRI (D_1)	DRI (D_2)
2.5	0.25	0.74 ± 0.08	0.53 ± 0.11	6.1	3.5	2.5	2.5	0.30 ± 0.10	0.35 ± 0.24	1.47	9.32
5	0.5	0.81 ± 0.06	0.77 ± 0.31	4.1	2.65	5	5	0.51 ± 0.02	0.57 ± 0.18	1.84	15.55
10	1	0.90 ± 0.04	0.99 ± 0.53			10	10	0.59 ± 0.03	0.59 ± 0.37	1.28	12.07
20	2	0.94 ± 0.02	1.28 ± 0.80			20	20	0.70 ± 0.01	0.90 ± 0.37	1.06	11.63
		(D_1) + (D_2) No constant ratio						(D_1) + (D_2) No constant ratio			
		Fractional inhibition	CI	DRI (D_1)	DRI (D_2)			Fractional inhibition	CI	DRI (D_1)	DRI (D_2)
1.25	0.1	0.21 ± 0.10	2.15 ± 0.35			1.25	2.5	0.26 ± 0.03	0.71 ± 0.03	2.39	7.10
1.25	0.25	0.44 ± 0.14	1.14 ± 0.30			1.25	5	0.28 ± 0.01	0.84 ± 0.11	2.66	4.08
2.5	0.1	0.26 ± 0.01	1.80 ± 0.70			2.5	2.5	0.43 ± 0.09	0.56 ± 0.13	2.64	20.00
2.5	0.25	0.64 ± 0.14	0.48 ± 0.17	4.35	2.16	2.5	5	0.44 ± 0.07	0.58 ± 0.02	2.75	10.61
2.5	0.5	0.74 ± 0.11	0.64 ± 0.14	6.18	1.75	2.5	10	0.45 ± 0.08	0.59 ± 0.13	2.87	5.60
5	0.1	0.40 ± 0.06	1.51 ± 0.45			5	2.5	0.50 ± 0.03	0.63 ± 0.11	1.76	29.45
5	0.25	0.75 ± 0.17	1.60 ± 0.45			5	5	0.50 ± 0.06	0.67 ± 0.13	1.76	14.72
5	0.5	0.79 ± 0.09	0.96 ± 0.31			5	10	0.54 ± 0.07	0.61 ± 0.13	2.08	9.15

Drugs were tested at different concentrations in order to find out which ratio yielded a better response. Based on the values of EC_{50} obtained from the dose-response curve of each drug, a first study was carried out in HCC1954 cells by mixing the two drugs (DHA and T-DM1) at various ratio from 50:1 to 10:1 (data not shown). The two-drug combination at a ratio of 10:1 yielded synergistic inhibition (Table 2). We have also extend the studies at a non-constant combination ratio (e.g., keep DHA at constant dose and vary the dose of T-DM1). DHA was used at concentration ranging from 1.25 to 5 μM (Table 2). Interestingly, the two-drug combination was more efficacious in inhibiting cell growth when cells were treated with DHA at 2.5 μM and T-DM1 at 0.25 μg/mL (Table 2, and Figure 4a, left panel).

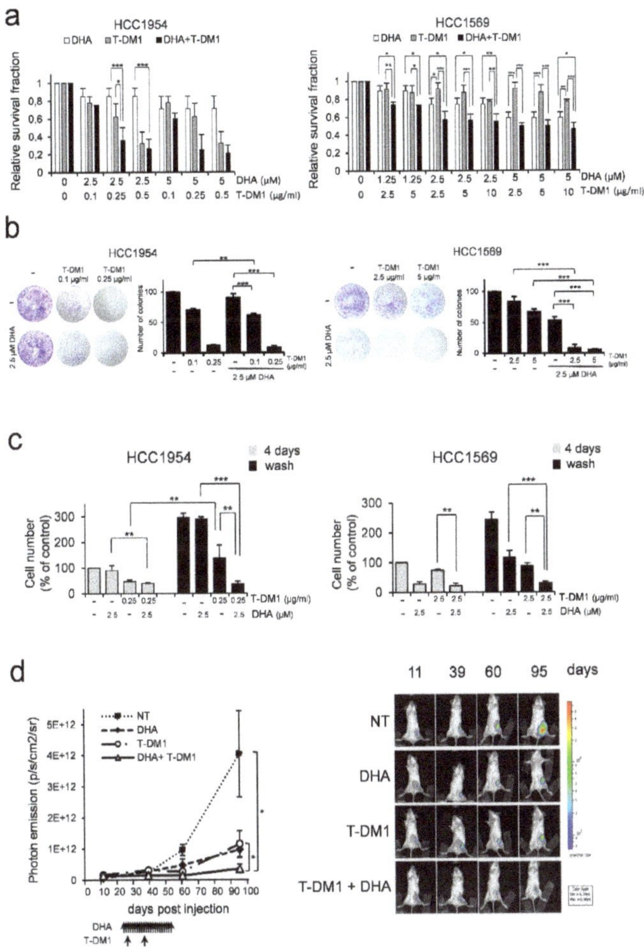

Figure 4. T-DM1 when combined with DHA is more effective in killing cancer cells. (**a**) Cell viability was assessed by ATP assay in cells pre-treated with DHA for 24 h and then treated with T-DM1 for 5 days. Relative survival fraction after treatment, control cells were set to "1". Values represent the mean ± SD, $n = 3$ * = $p < 0.05$, ** = $p < 0.01$, *** = $p < 0.001$ (**b**) Colony formation assay. Cells were treated as described in (a). A representative experiment of three with similar results is shown. Bar graphs show quantification analysis. Values represent the mean ± SD, $n = 3$. ** = $p < 0.01$, *** = $p < 0.001$. (**c**) Trypan blue exclusion assay was performed in cells pre-treated with DHA for 24 h and then treated with T-DM1 for 3 days (grey columns). On day 4, cells were washed with fresh media and further incubated for additional 4 days (black column). Data are expressed as the percentage of viable cells relative to controls. Values represent the mean ± SD, $n = 3$. * = $p < 0.05$, ** = $p < 0.01$, *** = $p < 0.001$. (**d**) Quantification of tumour xenografts growth by bioluminescence imaging. Luciferase-expressing HCC1954 cells were implanted into the mammary gland of CB17SCID mice. After 22 days animals were treated with: (1) vehicle; (2) DHA, administrated intraperitoneally (i.p.) 5 days per week, for 4 weeks, in a single daily dose of 25 mg/kg; (3) T-DM1, intravenously (i.v.), once every twelve days for a total of two injections at doses of 10 mg/kg; (4) the combination of DHA and T-DM1. Bioluminescence was detected in mice 11, 39, 60 and 95 days after tumour inoculation. Mean ± SEM is shown. Photon emission is measured as photons/sec/cm2/steradian (N = 5), $P < 0.05$, Mann Whitney test (left panel). Bioluminescence of a representative mouse at the indicated days after tumour inoculation. The intensity of light emission from the animals is represented in pseudo-colour scaling (right panel). * = $p < 0.05$.

The HCC1569 cell line was less responsive to T-DM1 treatment when compared to HCC1954 cells, as indicated by the EC_{50} value (EC_{50} =101.70 ± 44.8) reported in Table 2. We chose an arbitrary ratio of 1:1 of drug combinations in order to identify the minimum effective dose of each drug. Results obtained from drug combination studies are shown in Table 2 and Figure 4a (right panel). Overall, we found that DHA synergized with T-DM1, as indicated by the CI values less than 1 (Table 2). Interestingly, analysis of the dose reduction index (DRI) indicated that addition of DHA to T-DM1 allowed a dose-reduction for T-DM1 in both cell lines (Table 2). For instance, the addition of DHA at 2.5 μM to T-DM1 allowed up to 3-fold and ~10-fold reduction of T-DM1 in HCC1954 cells and HCC1569 cells, respectively.

The inhibition of cell growth induced by the combination of DHA and T-DM1 has been also confirmed by colony assay experiments in both cell lines (Figure 4b).

To further test the efficacy of the combination of DHA and T-DM1, we performed washout experiments. Cells were cultured at low density (1500 cells/cm^2) to maintain them in exponential growth. HCC1954 cells and HCC1569 cells were pre-treated with DHA at 2.5 μM and then treated with T-DM1 at 0.25 μg/mL and 2.5 μg/mL, respectively, for 3 days. This was followed by the removal of the drugs from the media and by a further incubation for additional 4 days. At the end of incubation time, the number of viable cells was determined by trypan blue dye exclusion assay. Figure 4c shows that a severe inhibition of cell growth was induced in both cell lines by the DHA and T-DM1 treatment. This effect persisted to a greater extent than those of each single agent upon removal of the drug-containing medium.

We then studied the inhibitory effects of DHA, T-DM1 and the two-drug combination in vivo. An orthotopic xenograft model of breast cancer was established by implantation of luciferase-expressing HCC1954 cells in the mouse mammary gland. Tumour cell engraftment and early detection of tumour growth was assessed by BLI analysis. To determine the in vivo antitumour activity of the treatments, pharmacological administration was initiated 22 days after tumour cell inoculation, when tumour bioluminescent emission reached an average value of 8×10^9 photons/sec/cm2/steradian (Supplementary Figure S3, panel a). Animals were divided into 4 experimental groups (N = 5 per group). Following treatment, mice were monitored for tumour recurrence. Within the period of follow-up, any single agent or the two-drug combination were well tolerated, with no signs of toxicity and weight loss (Supplementary Figure S3, panel b). As expected, T-DM1 treatment was effective in inhibiting tumour growth. However, the growth of the tumours resumed after having achieved a complete response (Figure 4d and Supplementary Figure S3, panel c), in line with data from literature [54,55]. In contrast, the combination DHA and T-DM1 was more efficient than each single agent on tumour inhibition throughout the observation period (Figure 4d, and Supplementary Figure S3, panel c), consistent with the results shown in Figure 4c.

Altogether, these data suggest that T-DM1 when combined with DHA is more effective in killing HER2+ BC cell lines.

3.5. *The Effects of Two-Drug Combination on HER2-Mediated Cell Signalling*

It has been reported that T-DM1 inhibits AKT phosphorylation in both trastuzumab-responsive and insensitive cell lines, suggesting that this effect could be mediated by the DM1 component of T-DM1 [56,57]. Therefore, we assessed the levels of activation of AKT in HCC1954 and HCC1569 cells pre-treated with DHA for 24 h and then exposed to T-DM1 for three days. Western blot analysis showed a great reduction of active AKT in cells treated with the combination of two drugs as compared to the effect induced by each single agent (Figure 5a). After four days of treatment, phospho-AKT levels were not affected by DHA (Figure 5a) and this suggests that only the T-DM1 and DHA combination was effective enough to reduce active AKT in the long-term period.

Then we investigated the effects of two-drug combination on the activation of p44/p42 MAPK (also called ERK1/2). Western blot analysis showed that the levels of phosphorylated ERK1/2 were barely reduced by the two-drug treatment.

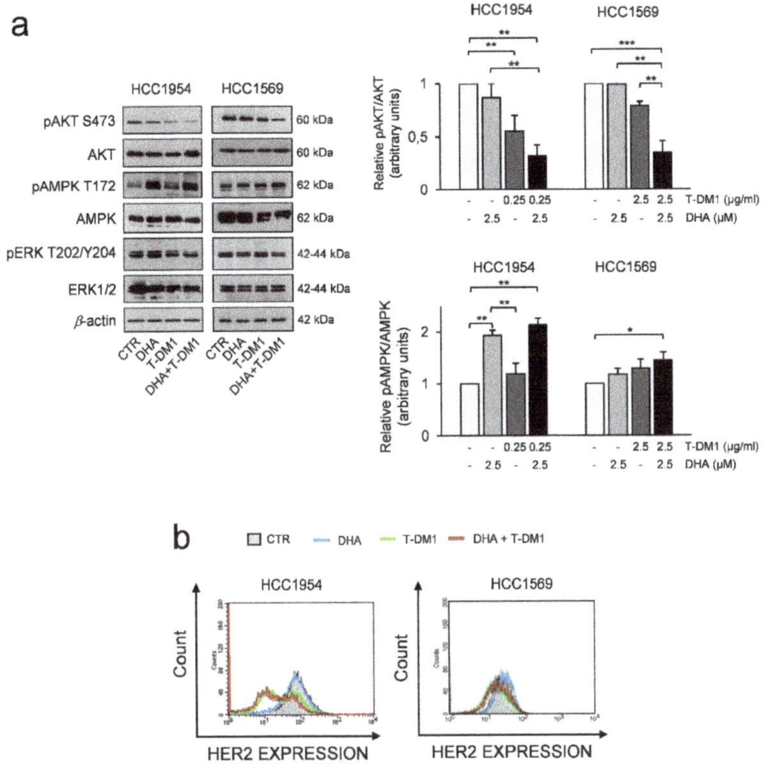

Figure 5. Effects of Two-drug combination on Protein kinase B (PKB/AKT hereafter referred as AKT), phospho-AMP-activated protein kinase (AMPK), and human epidermal growth factor receptor 2 (HER2) expression in HER2+ BC cell lines. (**a**) Western Blot analysis of the indicated proteins in cells pre-treated with DHA for 24 h and then treated with T-DM1 for 3 days. β-actin was used as loading control (right panel). For densitometric analysis the intensity of each band was normalized to the respective β-actin Quantification analysis was performed by using ImageJ software (left panel). (**b**) Quantitative flow cytometry (FACS) analysis of membrane HER2 expression. Cells were treated as described in (**a**) Histograms show one representative experiment of two with similar results. * = $p < 0.05$, ** = $p < 0.01$, *** = $p < 0.01$.

We also studied the phosphorylation of AMPK at its activation site Thr172. Interestingly, the two-drug treatment induced an increase of pphospho-AMPK, which was well-detectable in HCC1954 cells and, to a lesser degree, in HCC1569 cells when compared to control (Figure 5a). Interestingly, AMPK activation may play a role in mitosis and genomic stability beyond its role in the metabolic stress response [58,59] and could be activated upon T-DM1 or DHA treatment.

One critical point in the anti-HER2 therapy is that patients can experience altered HER2 status [60,61], and loss of HER2 expression could lead to a resistant phenotype. We found that T-DM1 treatment induced a reduction of cells with the highest levels of HER2 and an enrichment of cells with low levels of HER2 in HCC1954 cells, suggesting that T-DM1 eliminated cells expressing the highest HER2 levels, but still are responsive to the treatment, in line with data from the literature [9] (Figure 5b). In contrast, the HER2 status was not affected by T-DM1 in HCC1569 cells, which were less responsive to T-DM1 treatment (Figure 5b). Altogether, these data suggest that the inhibition of phospho-AKT could be mediated by the DM1 component of T-DM1 in HER2+ BC cell lines resistant to trastuzumab. A great reduction of active AKT is observed in these cells when treated with the combination of two drugs as compared to the effect induced by each single agent.

3.6. DHA in Combination with T-DM1 Led to Mitotic Catastrophe

Since T-DM1 is a microtubule-disrupting agent that may cause cell cycle arrest in the G2 and M phases, we evaluated its effects on cell cycle distribution when combined with DHA. The treatment of both cell lines with T-DM1 and with the DHA and T-DM1 combination inhibited cell-cycle progression, as indicated by the increased proportion of cells in G2/M phase (Figure 6a). Moreover, the treatment with T-DM1 induced a remarkable increase of cell population in sub-G1 area mainly in HCC1954 cells. The increase was even greater in HCC1954 cells exposed to the two-drug treatment, thus indicating apoptotic cells and/or fragmentation of DNA (Figure 6a).

An immunofluorescence analysis revealed aberrant mitotic spindles in both cell lines under all treatments. Both T-DM1 and the DHA and T-DM1 combination induced the formation of disorganized microtubule structures, unstructured tubulin foci, and multiple nuclei. In addition, a significant (approximately 2-fold) increase in density of microtubules was clearly found when HCC1569 cells were exposed to both DHA and DHA and T-DM1 combination, while we did not find any significant increase in density of microtubules in HCC1954 cells under all treatments (Supplementary Figure S4). Moreover, we also detected an aberrant distribution of phospho-TCTP in enlarged and morphologically altered HCC1569 and HCC1954 cells when treated with DHA and T-DM1 (Figure 6b).

Dividing cells with a defective mitotic apparatus undergo mitotic cell death/catastrophe [62]. Western blot analysis of cyclin B1 and phosphorylated histone H3 at Ser10 (p-Histone H3), two markers of M phase, showed that they were upregulated in HCC1569 cells, thus indicating that cells were arrested in mitosis. In addition, we found that activation of caspase 3 was induced by T-DM1 treatment. Notably, the increase of caspase-3 activity was even greater in cells exposed to DHA in combination with T-DM1, suggesting that cell death occurred in mitosis (Figure 6c). On the contrary, a different pattern was found in HCC1954 cells. Indeed, no accumulation of cyclin B1 was found in both T-DM1 and DHA and T-DM1 treated cells. However, the percentage of sub-G1 phase were significantly increased. To confirm the induction of apoptosis, we evaluated the active caspase 3 and the proteolytic cleavage of PARP. Figure 6c shows that T-DM1 treatment induced a significant increase of cell death as evaluated by the amount of cleaved PARP, which was further increased upon T-DM1 and DHA treatment. As DNA damage triggers PARP activation [63], we evaluated the amount of DNA damage levels by studying the increase in the phosphorylation levels of histone H2AX. Intriguingly, a greater DNA damage was found in HCC1954 cells subjected to two-drug combination than those exposed to any single agent, as indicated by the increased levels of γH2AX (Figure 6c).

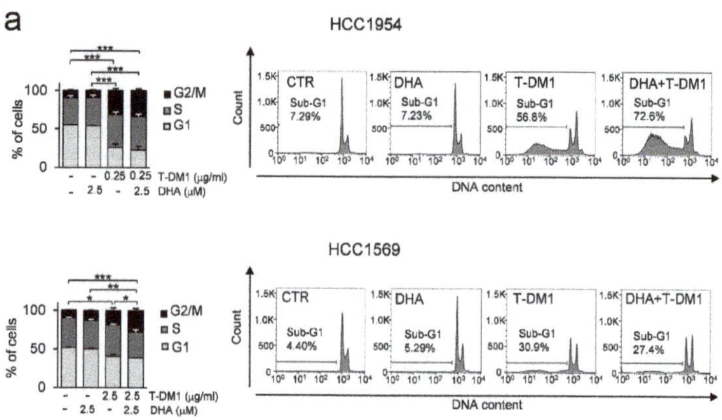

Figure 6. *Cont.*

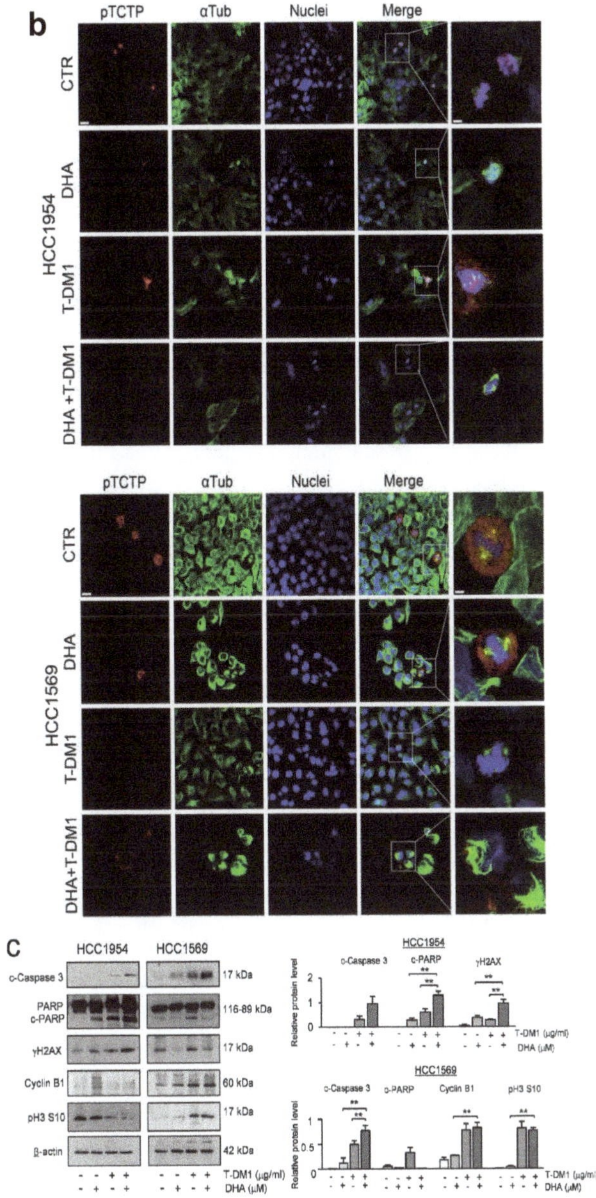

Figure 6. The combination of DHA and T-DM1 induces mitotic catastrophe in HER2+ BC cell lines (**a**) Cells were pre-treated with DHA for 24 h and then treated with T-DM1 for 3 days. The bar graphs show the distribution of cycling cells. The histograms show the percentage of Sub-G1 cells of one representative experiment. Values represent the mean ± SD, $n = 3$. * $p < 0.05$, ** $p < 0.01$, *** $p < 0.001$. (**b**) Cells were treated as described in (**a**). The overlay of the three fluorochromes is shown (Merge): phospho-TCTP (red) and αTubulin (green) and nuclei stained with Hoechst (blue), bar = 25 μm (left side). A greater detail of the boxed area is shown, bar = 5 μm (right side). Data from representative experiments are shown. $n = 2$. (**c**) Western Blot analysis of the indicated proteins in cell lysates of cells treated as described in a). β-actin was used as loading control (left panel). For densitometric analysis, the intensity of each band was normalized to the respective β-actin (right panel). * $p < 0.05$, ** $p < 0.01$.

Altogether, these data show that DHA in combination with T-DM1 induces severe mitotic defects and death in aggressive HER2+ BC cell lines.

4. Discussion

Microtubule-targeting drugs are extensively used in breast cancer therapy [64]. However, neurotoxicity and the development of resistance represent a clinical problem [65].

Spindle-targeting drugs, such as inhibitors of several kinases involved in the formation and function of the mitotic spindle, are in development to improve anti-mitotic chemotherapy [66]. PLK1 plays a crucial role in cell proliferation through its effects on chromosome segregation, spindle assembly, maturation of the centrosome, and cytokinesis during mitosis [27]. PLK1 is highly expressed in several human cancers with poor clinical outcome [67–69]. However, if PLK1 overexpression contributes to tumour formation, by inducing mitotic alteration and chromosomal instability, is still a matter of debate [70]. Beyond these conflicting results, PLK1 is currently studied as worthwhile therapeutic target [71]. Notably, the resistance to taxane therapy [71] or to T-DM1 treatment have been linked to PLK1 overexpression in HER2+ BC [72]. However, preclinical success with PLK1 inhibitors, has not translated well into clinical success. Dose-limiting toxicities of PLK1 inhibitors and specificity are critical issues [71].

In accordance with previously reported data, demonstrating TCTP's role as a key mitotic target of Plk1 in regulating anaphase progression [30] we recently, showed that TCTP is a direct substrate of PLK1 in mammary carcinoma cell lines. We also showed that DHA, by targeting phospho-TCTP, induces apoptosis and enhances the efficacy of chemotherapy and trastuzumab in HER2 overexpressing tumour cells [10].

The role of phospho-TCTP in cancer cells has not been thoroughly investigated, and only few studies in the literature show that the phosphorylation of TCTP is required for metaphase-anaphase transition [30,33,73]. In addition, high levels of phospho-TCTP are associated with adverse prognostic factors in breast cancer patients and in neuroblastoma patients, as previously reported by our group [10] and by Ramani et al [74], respectively. Moreover, we also showed that an increase of the nuclear phospho-TCTP level is associated with a poor clinical response to trastuzumab therapy in HER2-positive breast cancer [10].

In this study, we show that the reduction of phospho-TCTP levels induced by DHA produces a phenotype resembling almost that observed following PLK1 inhibition. As cellular models for our studies, we chose the HCC1954 and the HCC1569 cell lines, which resemble HER2 overexpressing tumours with *PI3KCA* mutation and loss of PTEN, respectively. In addition, both cells lines were resistant to trastuzumab therapy. Interestingly, malignant progression of HER2-positive breast cancer is often characterized by aberrant PI3K/AKT activation [1,75,76]. For instance, *PI3KCA* mutations and a high proliferation rate are unfavourable prognostic factors in relapsed and de novo metastatic HER2-positive breast cancers treated with trastuzumab [77]. In addition, the loss of at least one copy of the *PTEN* gene is associated with a poor worse outcome in HER2-positive breast cancer, although it is not yet clear whether it is predictive of trastuzumab resistance [78,79].

The levels of phospho-TCTP are critical during mitotic process. Indeed, MCF10A cells overexpressing the non-phosphorylatable form of TCTP showed numerous anomalies of the mitotic spindle. On the contrary, overexpression of the phosphorylatable form of protected MCF10A cells from aberrant mitosis. These data suggest that the reduction of phospho-TCTP levels could be deleterious for achieving proper cell division in growing conditions (Figure 3). In MCF10 cells overexpressing non-phosphorylatable form of TCTP we found high levels of aberrant mitosis. Abnormal mitosis in these cells occurred without any acquisition of DNA damage (Supplementary Figure S2), which in turn may help these cells to bypass surveillance mechanisms. In this context, we can speculate that overexpression of TCTP may protect from oxidative stress that could induce DNA damage in these cells, in accordance with our previously reported data that show TCTP as critical survival factor against oxidative stress [13]. It has been reported that in parasite *Brugia malayi*, TCTP protects DNA from oxidative damage [80]. However, we cannot exclude that DNA damage could occur during mitosis but it is likely to be transient or below a critical threshold necessary to activate p53 to levels that prevent proliferation. These findings have an important clinical value, as aggressive HER2+ BC cells with mutated p53 could be highly sensitive to the effect of DHA.

Remarkably, in line with data from preclinical and clinical studies showing that DHA has good safety profile [39,41,42,51], we also showed that the EC_{50} of DHA was significantly lower in cancer cell lines (<10 μM) than in non-tumourigenic epithelial cells MCF10A (>30 μM) (Table 1). This suggests that DHA has little side effects, thus representing a potentially interesting approach for breast cancer therapy.

DHA is a pro-oxidant agent [53] that increases the cellular levels of ROS and leads to DNA damage, in line with the recent observation that artesunate, whose active metabolite is DHA, induces oxidative DNA lesions and DNA double-strand breaks (DSB) [81,82] (Figure 2). Altogether, these findings have some important implications. First, the reduction of TCTP could contribute to DHA-mediated oxidative damage through the increase of oxidative stress, in line with our previously reported findings [13]. The second is that a critical threshold of DSBs is necessary to prevent mitotic entry [83]. In this context, we can speculate that, in HCC1954 cells, DHA induces DNA damage at an extent below this critical threshold, Therefore, cells can still enter mitosis; thus, contributing to increase genome instability. However, excessive genomic instability has a deleterious effect to the viability of cancer cells, and could be explored for therapeutic purposes. Moreover, the increase of ROS levels in *PTEN*-deficient HCC1569 cells or *PIK3CA* mutant-HCC1954 cells could be their Achilles' heel, since it has been reported that high levels of active AKT could increase the susceptibility of cancer cells to oxidative stress [84]. PI3K signalling plays a critical role in cell proliferation, mainly through phosphorylation of AKT. In this context, we can speculate that DHA induces metabolic perturbation through the reduction of AKT phosphorylation levels (Figure 2). This effect is further enhanced by T-DM1 (Figure 5). Beyond the induction of severe mitotic perturbations, the inhibition of pAKT by the treatment may contribute to the successful outcomes of DHA and T-DM1 therapy. These data may have clinical relevance since molecular alterations involving the PI3K/AKT pathway are frequently observed in advanced HER2+ BC.

By reducing phospho-TCTP levels, DHA induces both mitotic aberration and the formation of disorganized microtubule structures (Figure 1; Figure 3). Therefore, DHA exacerbated the cytotoxicity of T-DM1 as inhibitor of microtubules. Moreover, analysis of the dose reduction index (DRI) for each drug in their combination indicated that addition of DHA to T-DM1 allowed a dose-reduction for T-DM1 in both cell lines. These data have important clinical implication, as it has been reported that patients treated with T-DM1 have more adverse events than those treated with adjuvant trastuzumab [6] (Table 2).

The combination treatment was more effective in killing breast cancer cells when compared to the effect induced by any single agent. Notably, the in vivo data show that the growth of tumour cells resumed after having achieved a complete response to T-DM1 treatment. Conversely, DHA and T-DM1 treatment improved the long-term efficacy of T-DM1 and induced a severe and irreversible cytotoxic effect, even after treatment interruption (Figure 4; Figure 6).

Collectively, these results suggest that DHA synergizes with T-DM1 in trastuzumab-resistant cell lines. By reducing phospho-TCTP levels, DHA enhances the effects of T-DM1 by increasing its effectiveness as a poison for microtubules, and in turn, may cause a dramatic inhibition of cell growth in aggressive breast cancer resistant to trastuzumab therapy. This new therapeutic protocol could lead to a clinically significant improvement in the response to T-DM1 therapy.

5. Conclusions

In our study, we showed that phospho-TCTP is essential for correct mitosis. Reduction of phospho-TCTP levels by DHA causes mitotic aberration and increases microtubule density in trastuzumab-resistant breast cancer cells. In this way, DHA enhances the effect of T-DM1 as a poison for microtubules, leading tumour cells into an unstable state no longer compatible with viability.

T-DM1 is already an FDA-approved agent for advanced breast cancer. DHA is an antimalarial drug, already tested for its toxicity, and has been approved by the European Medicines Agency (EMA) for human use. Due to its low cost and high tolerability profile, it is currently undergoing clinical trials in various types of neoplasia. This study provides the rationale for the design of a new clinical protocol.

Supplementary Materials: The following are available online at http://www.mdpi.com/2073-4409/9/5/1260/s1, Figure S1: Effect of trastuzumab and pertuzumab on cell growth of HER2+ BC cell lines, Figure S2: DHA induces an increase of ROS levels in MCF10A-pBabe, MCF10A-AATCTP and MCF10A-WTTCTP cells, Figure S3: In vivo efficacy of DHA with T-PM1 in HCC1954 xenografts, Figure S4: Effect of two-drug combination on microtubule density in HCC1569 and HCC1954 cells.

Author Contributions: F.d.B. and M.L. contributed equally to this work. S.D., E.K.K. contributed equally to this work. Conceptualization, F.d.B. and M.L.; methodology, M.L., F.d.B., E.K.K., S.D.; validation, M.L., S.D., E.E.K., G.P., I.M., G.T.; formal analysis, M.L., S.D., E.K.K., I.M., G.P.; investigation, M.L., S.D., E.K.K., A.G., M.R., G.P.; I.M., G.T., S.B., F.D.G.; resources, M.L., G.P.; writing—original draft preparation, M.L.; writing—review and editing, M.L., F.d.B., F.B., C.V., G.P., A.G.; visualization, M.L., S.D., E.K.K., I.M.; supervision, M.L., F.d.B.; project administration, M.L.; funding acquisition, M.L. All authors have read and agreed to the published version of the manuscript.

Funding: This research was supported by Roche S.p.A. The sponsor had no role in the design, execution, interpretation, or writing of the study.

Acknowledgments: We thank Daniela Barilà for the helpful reading and suggestions on our manuscript, Antonio Rossi for assistance in retroviral transduction experiments, and Mauro Fabiani for technical support. Special thanks to Genentech, USA Inc. for providing us with T-DM1 under Material Transfer Agreement with MTA number OR-129895.

Conflicts of Interest: The authors declare no conflicts of interest.

References

1. Arteaga, C.L.; Engelman, J.A. ERBB receptors: From oncogene discovery to basic science to mechanism-based cancer therapeutics. *Cancer Cell* **2014**, *25*, 282–303. [CrossRef] [PubMed]
2. Larionov, A. Current Therapies for Human Epidermal Growth Factor Receptor 2-Positive Metastatic Breast Cancer Patients. *Front. Oncol.* **2018**, *8*, 89. [CrossRef] [PubMed]
3. Vernieri, C.; Milano, M.; Brambilla, M.; Mennitto, A.; Maggi, C.; Cona, M.S.; Prisciandaro, M.; Fabbroni, C.; Celio, L.; Mariani, G.; et al. Resistance mechanisms to anti-HER2 therapies in HER2-positive breast cancer: Current knowledge, new research directions and therapeutic perspectives. *Crit. Rev. Oncol.* **2019**, *139*, 53–66. [CrossRef] [PubMed]
4. Yan, H.; Yu, K.; Zhang, K.; Liu, L.; Li, Y. Efficacy and safety of trastuzumab emtansine (T-DM1) in the treatment of HER2-positive metastatic breast cancer (MBC): A meta-analysis of randomized controlled trial. *Oncotarget* **2017**, *8*, 102458–102467. [CrossRef]
5. Diéras, V.; Miles, D.; Verma, S.; Pegram, M.; Welslau, M.; Baselga, J.; E Krop, I.; Blackwell, K.; Hoersch, S.; Xu, J.; et al. Trastuzumab emtansine versus capecitabine plus lapatinib in patients with previously treated HER2-positive advanced breast cancer (EMILIA): A descriptive analysis of final overall survival results from a randomised, open-label, phase 3 trial. *Lancet Oncol.* **2017**, *18*, 732–742. [CrossRef]
6. Von Minckwitz, G.; Huang, C.-S.; Mano, M.S.; Loibl, S.; Mamounas, E.P.; Untch, M.; Wolmark, N.; Rastogi, P.; Schneeweiss, A.; Redondo, A.; et al. Trastuzumab Emtansine for Residual Invasive HER2-Positive Breast Cancer. *N. Engl. J. Med.* **2018**, *380*, 617–628. [CrossRef]
7. Von Minckwitz, G.; Procter, M.; De Azambuja, E.; Zardavas, D.; Benyunes, M.; Viale, G.; Suter, T.; Arahmani, A.; Rouchet, N.; Clark, E.; et al. Adjuvant Pertuzumab and Trastuzumab in Early HER2-Positive Breast Cancer. *N. Engl. J. Med.* **2017**, *377*, 122–131. [CrossRef]
8. García-Alonso, S.; Ocana, A.; Pandiella, A. Resistance to Antibody–Drug Conjugates. *Cancer Res.* **2018**, *78*, 2159–2165. [CrossRef]
9. Li, G.; Guo, J.; Shen, B.-Q.; Yadav, D.B.; Sliwkowski, M.X.; Crocker, L.M.; Lacap, J.A.; Phillips, G.D.L. Mechanisms of Acquired Resistance to Trastuzumab Emtansine in Breast Cancer Cells. *Mol. Cancer Ther.* **2018**, *17*, 1441–1453. [CrossRef]

10. Lucibello, M.; Adanti, S.; Antelmi, E.; Dezi, D.; Ciafre, S.; Carcangiu, M.L.; Zonfrillo, M.; Nicotera, G.; Sica, L.; De Braud, F.; et al. Phospho-TCTP as a therapeutic target of dihydroartemisinin for aggressive breast cancer cells. *Oncotarget* **2015**, *6*, 5275–5291. [CrossRef]
11. Bommer, U. The Translational Controlled Tumour Protein TCTP: Biological Functions and Regulation. *Results Probl. Cell Differ.* **2017**, *64*, 69–126. [CrossRef]
12. Chen, S.H.; Wu, P.-S.; Chou, C.-H.; Yan, Y.-T.; Liu, H.; Weng, S.-Y.; Yang-Yen, H.-F. A Knockout Mouse Approach Reveals that TCTP Functions as an Essential Factor for Cell Proliferation and Survival in a Tissue- or Cell Type–specific Manner. *Mol. Biol. Cell* **2007**, *18*, 2525–2532. [CrossRef] [PubMed]
13. Lucibello, M.; Gambacurta, A.; Zonfrillo, M.; Pierimarchi, P.; Serafino, A.; Rasi, G.; Rubartelli, A.; Garaci, E. TCTP is a critical survival factor that protects cancer cells from oxidative stress-induced cell-death. *Exp. Cell Res.* **2011**, *317*, 2479–2489. [CrossRef] [PubMed]
14. Amson, R.; Karp, J.E.; Telerman, A. Lessons from tumor reversion for cancer treatment. *Curr. Opin. Oncol.* **2013**, *25*, 59–65. [CrossRef]
15. Tuynder, M.; Susini, L.; Prieur, S.; Besse, S.; Fiucci, G.; Amson, R.; Telerman, A. Biological models and genes of tumor reversion: Cellular reprogramming through tpt1/TCTP and SIAH-1. *Proc. Natl. Acad. Sci. USA* **2002**, *99*, 14976–14981. [CrossRef]
16. Hsu, Y.-C.; Chern, J.J.; Cai, Y.; Liu, M.; Choi, K.-W. Drosophila TCTP is essential for growth and proliferation through regulation of dRheb GTPase. *Nature* **2007**, *445*, 785–788. [CrossRef]
17. Tuynder, M.; Fiucci, G.; Prieur, S.; Lespagnol, A.; Géant, A.; Beaucourt, S.; Duflaut, D.; Besse, S.; Susini, L.; Cavarelli, J.; et al. Translationally controlled tumor protein is a target of tumor reversion. *Proc. Natl. Acad. Sci. USA* **2004**, *101*, 15364–15369. [CrossRef]
18. Amson, R.; Pece, S.; Marine, J.-C.; Di Fiore, P.P.; Telerman, A. TPT1/ TCTP-regulated pathways in phenotypic reprogramming. *Trends Cell Biol.* **2013**, *23*, 37–46. [CrossRef]
19. Amson, R.; Auclair, C.; Andre, F.; Karp, J.; Telerman, A. Targeting TCTP with Sertraline and Thioridazine in Cancer Treatment. *Results Probl. Cell Differ.* **2017**, *64*, 283–290. [CrossRef]
20. Amson, R.; Pece, S.; Lespagnol, A.; Vyas, R.; Mazzarol, G.; Tosoni, D.; Colaluca, I.; Viale, G.; Rodrigues-Ferreira, S.; Wynendaele, J.; et al. Reciprocal repression between P53 and TCTP. *Nat. Med.* **2011**, *18*, 91–99. [CrossRef]
21. Bae, S.-Y.; Kim, H.J.; Lee, K.-J.; Lee, K. Translationally Controlled Tumor Protein induces epithelial to mesenchymal transition and promotes cell migration, invasion and metastasis. *Sci. Rep.* **2015**, *5*, 8061. [CrossRef] [PubMed]
22. Mishra, D.K.; Srivastava, P.; Sharma, A.; Prasad, R.; Bhuyan, S.K.; Malage, R.; Kumar, P.; Yadava, P. Translationally controlled tumor protein (TCTP) is required for TGF-β1 induced epithelial to mesenchymal transition and influences cytoskeletal reorganization. *Biochim. et Biophys. Acta (BBA) - Bioenerg.* **2018**, *1865*, 67–75. [CrossRef] [PubMed]
23. Miao, X.; Chen, Y.-B.; Xu, S.-L.; Zhao, T.; Liu, J.-Y.; Li, Y.-R.; Wang, J.; Zhang, J.; Guo, G.-Z. TCTP overexpression is associated with the development and progression of glioma. *Tumor Biol.* **2013**, *34*, 3357–3361. [CrossRef] [PubMed]
24. Chen, C.; Deng, Y.; Hua, M.; Xi, Q.; Liu, R.; Yang, S.; Liu, J.; Zhong, J.; Tang, M.; Lu, S.; et al. Expression and clinical role of TCTP in epithelial ovarian cancer. *J. Mol. Histol.* **2015**, *46*, 145–156. [CrossRef] [PubMed]
25. Baylot, V.; Karaki, S.; Rocchi, P. TCTP Has a Crucial Role in the Different Stages of Prostate Cancer Malignant Progression. *Results Probl. Cell Differ.* **2017**, *64*, 255–261. [CrossRef]
26. Giusiano, S.; Garcia, S.; Andrieu, C.; Dusetti, N.; Bastide, C.; Gleave, M.; Iovanna, J.; Rocchi, P.; Taranger-Charpin, C. TP53INP1 overexpression in prostate cancer correlates with poor prognostic factors and is predictive of biological cancer relapse. *Prostate* **2011**, *72*, 117–128. [CrossRef]
27. Combes, G.; Alharbi, I.; Braga, L.G.; Elowe, S. Playing polo during mitosis: PLK1 takes the lead. *Oncogene* **2017**, *36*, 4819–4827. [CrossRef]
28. Li, S.; Ge, F. Current Understanding of the TCTP Interactome. *Results Probl. Cell Differ.* **2017**, *64*, 127–136. [CrossRef]
29. Liu, H.; Peng, H.-W.; Cheng, Y.-S.; Yuan, H.S.; Yang-Yen, H.-F. Stabilization and Enhancement of the Antiapoptotic Activity of Mcl-1 by TCTP. *Mol. Cell. Biol.* **2005**, *25*, 3117–3126. [CrossRef]
30. Yarm, F.R. Plk Phosphorylation Regulates the Microtubule-Stabilizing Protein TCTP. *Mol. Cell. Biol.* **2002**, *22*, 6209–6221. [CrossRef]

31. Cucchi, U.; Gianellini, L.M.; De Ponti, A.; Sola, F.; Alzani, R.; Patton, V.; Pezzoni, A.; Troiani, S.; Saccardo, M.B.; Rizzi, S.; et al. Phosphorylation of TCTP as a marker for polo-like kinase-1 activity in vivo. *Anticancer. Res.* **2010**, *30*, 4973–4985.
32. Lemmens, B.; Hégarat, N.; Akopyan, K.; Sala-Gaston, J.; Bartek, J.; Hochegger, H.; Lindqvist, A. DNA Replication Determines Timing of Mitosis by Restricting CDK1 and PLK1 Activation. *Mol. Cell* **2018**, *71*, 117–128.e3. [CrossRef] [PubMed]
33. Jeon, H.-J.; You, S.Y.; Park, Y.S.; Chang, J.W.; Kim, J.-S.; Oh, J.S. TCTP regulates spindle microtubule dynamics by stabilizing polar microtubules during mouse oocyte meiosis. *Biochim. et Biophys. Acta (BBA) - Bioenerg.* **2016**, *1863*, 630–637. [CrossRef] [PubMed]
34. Gachet, Y.; Tournier, S.; Lee, M.; Lazaris-Karatzas, A.; Poulton, T.; A Bommer, U. The growth-related, translationally controlled protein P23 has properties of a tubulin binding protein and associates transiently with microtubules during the cell cycle. *J. Cell Sci.* **1999**, *112*, 1257–1271. [PubMed]
35. Bhisutthibhan, J.; Meshnick, S.R. Immunoprecipitation of [3H]Dihydroartemisinin Translationally Controlled Tumor Protein (TCTP) Adducts from Plasmodium falciparum-Infected Erythrocytes by Using Anti-TCTP Antibodies. *Antimicrob. Agents Chemother.* **2001**, *45*, 2397–2399. [CrossRef]
36. Efferth, T. Mechanistic perspectives for 1,2,4-trioxanes in anti-cancer therapy. *Drug Resist. Updat.* **2005**, *8*, 85–97. [CrossRef]
37. Zhang, F.; Ma, Q.; Xu, Z.; Liang, H.; Li, H.; Ye, Y.; Xiang, S.; Zhang, Y.; Jiang, L.; Hu, Y.; et al. Dihydroartemisinin inhibits TCTP-dependent metastasis in gallbladder cancer. *J. Exp. Clin. Cancer Res.* **2017**, *36*, 68. [CrossRef]
38. Tu, Y. The discovery of artemisinin (qinghaosu) and gifts from Chinese medicine. *Nat. Med.* **2011**, *17*, 1217–1220. [CrossRef]
39. Jansen, F.H.; Adoubi, I.; C, K.C.J.; De Cnodder, T.; Jansen, N.; Tschulakow, A.; Efferth, T. First study of oral Artenimol-R in advanced cervical cancer: Clinical benefit, tolerability and tumor markers. *Anticancer. Res.* **2011**, *31*, 4417–4422.
40. Ericsson, T.; Blank, A.; Von Hagens, C.; Ashton, M.; Äbelö, A. Population pharmacokinetics of artesunate and dihydroartemisinin during long-term oral administration of artesunate to patients with metastatic breast cancer. *Eur. J. Clin. Pharmacol.* **2014**, *70*, 1453–1463. [CrossRef]
41. Von Hagens, C.; Walter-Sack, I.; Goeckenjan, M.; Osburg, J.; Storch-Hagenlocher, B.; Sertel, S.; Elsässer, M.; Remppis, B.A.; Edler, L.; Munzinger, J.; et al. Prospective open uncontrolled phase I study to define a well-tolerated dose of oral artesunate as add-on therapy in patients with metastatic breast cancer (ARTIC M33/2). *Breast Cancer Res. Treat.* **2017**, *164*, 359–369. [CrossRef] [PubMed]
42. A Morris, C.; Duparc, S.; Borghini-Fuhrer, I.; Jung, D.; Shin, C.-S.; Fleckenstein, L. Review of the clinical pharmacokinetics of artesunate and its active metabolite dihydroartemisinin following intravenous, intramuscular, oral or rectal administration. *Malar. J.* **2011**, *10*, 263. [CrossRef] [PubMed]
43. Li, Q.; Remich, S.; Miller, R.S.; Ogutu, B.; Otieno, W.; Melendez, V.; Teja-Isavadharm, P.; Weina, P.J.; Hickman, M.; Smith, B.; et al. Pharmacokinetic evaluation of intravenous artesunate in adults with uncomplicated falciparum malaria in Kenya: A phase II study. *Malar. J.* **2014**, *13*, 281. [CrossRef] [PubMed]
44. Efferth, T.; Li, P.C.; Konkimalla, B.; Kaina, B. From traditional Chinese medicine to rational cancer therapy. *Trends Mol. Med.* **2007**, *13*, 353–361. [CrossRef]
45. Chou, T.-C. Drug Combination Studies and Their Synergy Quantification Using the Chou-Talalay Method. *Cancer Res.* **2010**, *70*, 440–446. [CrossRef]
46. Di Rocco, G.; Gentile, A.; Antonini, A.; Truffa, S.; Piaggio, G.; Capogrossi, M.C.; Toietta, G. Analysis of Biodistribution and Engraftment into the Liver of Genetically Modified Mesenchymal Stromal Cells Derived from Adipose Tissue. *Cell Transplant.* **2012**, *21*, 1997–2008. [CrossRef]
47. Holen, I.; Speirs, V.; Morrissey, B.; Blyth, K. In vivo models in breast cancer research: Progress, challenges and future directions. *Dis. Model. Mech.* **2017**, *10*, 359–371. [CrossRef]
48. Goel, S.; Krop, I.E. Deciphering the Role of Phosphatidylinositol 3-Kinase Mutations in Human Epidermal Growth Factor Receptor 2–Positive Breast Cancer. *J. Clin. Oncol.* **2015**, *33*, 1407–1409. [CrossRef]
49. Berns, K.; Horlings, H.; Halfwerk, J.; Hennessy, B.; Linn, S.; Hauptmann, M.; Mills, G.; Van De Vijver, M.; Bernards, R. 57 A functional genetic approach identifies the PI3K pathway as a major determinant of Trastuzumab resistance in breast cancer. *Eur. J. Cancer Suppl.* **2009**, *7*, 17. [CrossRef]

50. O'Brien, N.A.; Browne, B.C.; Chow, L.; Wang, Y.; Ginther, C.; Arboleda, J.; Duffy, M.J.; Crown, J.; O'Donovan, N.; Slamon, D.J. Activated phosphoinositide 3-kinase/AKT signaling confers resistance to trastuzumab but not lapatinib. *Mol. Cancer Ther.* **2010**, *9*, 1489–1502. [CrossRef]
51. Tan, B.; Fleckenstein, L.; Yu, K.S.; Jang, I.J. Population Pharmacokinetics of Artesunate and Dihydroartemisinin in Healthy Volunteers. *Am. J. Trop. Med. Hyg.* **2008**, *79*, 135.
52. Valsasina, B.; Beria, I.; Alli, C.; Alzani, R.; Avanzi, N.; Ballinari, D.; Cappella, P.; Caruso, M.; Casolaro, A.; Ciavolella, A.; et al. NMS-P937, an Orally Available, Specific Small-Molecule Polo-like Kinase 1 Inhibitor with Antitumor Activity in Solid and Hematologic Malignancies. *Mol. Cancer Ther.* **2012**, *11*, 1006–1016. [CrossRef] [PubMed]
53. Zhang, S.; Gerhard, G.S. Heme Mediates Cytotoxicity from Artemisinin and Serves as a General Anti-Proliferation Target. *PLoS ONE* **2009**, *4*, e7472. [CrossRef] [PubMed]
54. Jumbe, N.L.; Xin, Y.; Leipold, D.D.; Crocker, L.; Dugger, D.; Mai, E.; Sliwkowski, M.X.; Fielder, P.J.; Tibbitts, J. Modeling the efficacy of trastuzumab-DM1, an antibody drug conjugate, in mice. *J. Pharmacokinet. Pharmacodyn.* **2010**, *37*, 221–242. [CrossRef]
55. Schwarz, L.; Hutchinson, K.E.; Rexer, B.N.; Estrada, M.V.; Gonzalez-Ericsson, P.I.; Sanders, M.E.; Dugger, T.C.; Formisano, L.; Guerrero-Zotano, A.; Red-Brewer, M.; et al. An ERBB1-3 Neutralizing Antibody Mixture With High Activity Against Drug-Resistant HER2+ Breast Cancers With ERBB Ligand Overexpression. *J. Natl. Cancer Inst.* **2017**, *109*. [CrossRef]
56. Junttila, T.T.; Akita, R.W.; Parsons, K.; Fields, C.; Phillips, G.D.L.; Friedman, L.S.; Sampath, D.; Sliwkowski, M.X. Ligand-Independent HER2/HER3/PI3K Complex Is Disrupted by Trastuzumab and Is Effectively Inhibited by the PI3K Inhibitor GDC-0941. *Cancer Cell* **2011**, *20*, 818. [CrossRef]
57. Phillips, G.D.L.; Fields, C.T.; Li, G.; Dowbenko, D.; Schaefer, G.; Miller, K.; Andre, F.; Burris, H.A.; Albain, K.S.; Harbeck, N.; et al. Dual Targeting of HER2-Positive Cancer with Trastuzumab Emtansine and Pertuzumab: Critical Role for Neuregulin Blockade in Antitumor Response to Combination Therapy. *Clin. Cancer Res.* **2013**, *20*, 456–468. [CrossRef]
58. Zadra, G.; Batista, J.L.; Loda, M. Dissecting the Dual Role of AMPK in Cancer: From Experimental to Human Studies. *Mol. Cancer Res.* **2015**, *13*, 1059–1072. [CrossRef]
59. Shen, Y.; Sherman, J.W.; Chen, X.; Wang, R. Phosphorylation of CDC25C by AMP-activated protein kinase mediates a metabolic checkpoint during cell-cycle G2/M-phase transition. *J. Biol. Chem.* **2018**, *293*, 5185–5199. [CrossRef]
60. Lindström, L.S.; Karlsson, E.; Wilking, U.; Johansson, U.; Hartman, J.; Lidbrink, E.; Hatschek, T.; Skoog, L.; Bergh, J. Clinically Used Breast Cancer Markers Such As Estrogen Receptor, Progesterone Receptor, and Human Epidermal Growth Factor Receptor 2 Are Unstable Throughout Tumor Progression. *J. Clin. Oncol.* **2012**, *30*, 2601–2608. [CrossRef]
61. Yoshida, A.; Hayashi, N.; Suzuki, K.; Takimoto, M.; Nakamura, S.; Yamauchi, H. Change in HER2 status after neoadjuvant chemotherapy and the prognostic impact in patients with primary breast cancer. *J. Surg. Oncol.* **2017**, *116*, 1021–1028. [CrossRef] [PubMed]
62. Mc Gee, M.M. Targeting the Mitotic Catastrophe Signaling Pathway in Cancer. *Mediat. Inflamm.* **2015**, *2015*, 1–13. [CrossRef] [PubMed]
63. Bai, P. Biology of Poly(ADP-Ribose) Polymerases: The Factotums of Cell Maintenance. *Mol. Cell* **2015**, *58*, 947–958. [CrossRef] [PubMed]
64. Jordan, M.A.; Wilson, L. Microtubules as a target for anticancer drugs. *Nat. Rev. Cancer* **2004**, *4*, 253–265. [CrossRef] [PubMed]
65. Rowinsky, E.K.; Eisenhauer, E.A.; Chaudhry, V.; Arbuck, S.G.; Donehower, R.C. Clinical toxicities encountered with paclitaxel (Taxol). *Semin. Oncol.* **1993**, *20*, 1–15.
66. Shi, J.; Mitchison, T.J. Cell death response to anti-mitotic drug treatment in cell culture, mouse tumor model and the clinic. *Endocrine-Related Cancer* **2017**, *24*, T83–T96. [CrossRef]
67. Bhola, N.E.; Jansen, V.M.; Bafna, S.; Giltnane, J.M.; Balko, J.M.; Estrada, M.V.; Meszoely, I.; Mayer, I.; Abramson, V.; Ye, F.; et al. Kinome-wide Functional Screen Identifies Role of PLK1 in Hormone-Independent, ER-Positive Breast Cancer. *Cancer Res.* **2015**, *75*, 405–414. [CrossRef]
68. Degenhardt, Y.; Lampkin, T. Targeting Polo-like Kinase in Cancer Therapy. *Clin. Cancer Res.* **2010**, *16*, 384–389. [CrossRef]

69. Watanabe, G.; Ishida, T.; Furuta, A.; Takahashi, S.; Watanabe, M.; Nakata, H.; Kato, S.; Ishioka, C.; Ohuchi, N. Combined Immunohistochemistry of PLK1, p21, and p53 for Predicting TP53 Status. *Am. J. Surg. Pathol.* **2015**, *39*, 1026–1034. [CrossRef]
70. De Cárcer, G.; Venkateswaran, S.V.; Salgueiro, L.; El Bakkali, A.; Somogyi, K.; Rowald, K.; Montañés, P.; Sanclemente, M.; Escobar, B.; De Martino, A.; et al. Plk1 overexpression induces chromosomal instability and suppresses tumor development. *Nat. Commun.* **2018**, *9*, 3012. [CrossRef]
71. Gutteridge, R.E.A.; Ndiaye, M.A.; Liu, X.; Ahmad, N. Plk1 Inhibitors in Cancer Therapy: From Laboratory to Clinics. *Mol. Cancer Ther.* **2016**, *15*, 1427–1435. [CrossRef] [PubMed]
72. Saatci, Ö.; Borgoni, S.; Akbulut, O.; Durmus, S.; Raza, U.; Eyüpoğlu, E.; Alkan, C.; Akyol, A.; Kutuk, O.; Wiemann, S.; et al. Targeting PLK1 overcomes T-DM1 resistance via CDK1-dependent phosphorylation and inactivation of Bcl-2/xL in HER2-positive breast cancer. *Oncogene* **2018**, *37*, 2251–2269. [CrossRef] [PubMed]
73. Jaglarz, M.K.; Bazile, F.; Laskowska, K.; Polański, Z.; Chesnel, F.; Borsuk, E.; Kloc, M.; Kubiak, J.Z. Association of TCTP with Centrosome and Microtubules. *Biochem. Res. Int.* **2012**, *2012*, 1–8. [CrossRef] [PubMed]
74. Ramani, P.; Nash, R.; Sowa-Avugrah, E.; Rogers, C. High levels of polo-like kinase 1 and phosphorylated translationally controlled tumor protein indicate poor prognosis in neuroblastomas. *J. Neuro-Oncol.* **2015**, *125*, 103–111. [CrossRef] [PubMed]
75. Stemke-Hale, K.; Gonzalez-Angulo, A.M.; Lluch, A.; Neve, R.M.; Kuo, W.-L.; Davies, M.; Carey, M.; Hu, Z.; Guan, Y.; Sahin, A.; et al. An integrative genomic and proteomic analysis of PIK3CA, PTEN, and AKT mutations in breast cancer. *Cancer Res.* **2008**, *68*, 6084–6091. [CrossRef] [PubMed]
76. Seo, Y.; Park, Y.H.; Ahn, J.S.; Im, Y.-H.; Nam, S.J.; Cho, S.Y.; Cho, E.Y. PIK3CA Mutations and Neoadjuvant Therapy Outcome in Patients with Human Epidermal Growth Factor Receptor 2-Positive Breast Cancer: A Sequential Analysis. *J. Breast Cancer* **2018**, *21*, 382–390. [CrossRef]
77. Kotoula, V.; Tsakiri, K.; Koliou, G.-A.; Lazaridis, G.; Papadopoulou, K.; Giannoulatou, E.; Tikas, I.; Christodoulou, C.; Chatzopoulos, K.; Bobos, M.; et al. Relapsed and De Novo Metastatic HER2-positive Breast Cancer Treated With Trastuzumab: Tumor Genotypes and Clinical Measures Associated With Patient Outcome. *Clin. Breast Cancer* **2019**, *19*, 113–125.e4. [CrossRef]
78. Stern, H.M.; Gardner, H.; Burzykowski, T.; Elatre, W.; O'Brien, C.; Lackner, M.; Pestano, G.A.; Santiago, A.; Villalobos, I.; Eiermann, W.; et al. PTEN Loss Is Associated with Worse Outcome in HER2-Amplified Breast Cancer Patients but Is Not Associated with Trastuzumab Resistance. *Clin. Cancer Res.* **2015**, *21*, 2065–2074. [CrossRef]
79. Tang, N.; Marshall, W.F. Centrosome positioning in vertebrate development. *J. Cell Sci.* **2012**, *125*, 4951–4961. [CrossRef]
80. Gnanasekar, M.; Ramaswamy, K. Translationally controlled tumor protein of Brugia malayi functions as an antioxidant protein. *Parasitol. Res.* **2007**, *101*, 1533–1540. [CrossRef]
81. Berdelle, N.; Nikolova, T.; Quirós, S.; Efferth, T.; Kaina, B. Artesunate Induces Oxidative DNA Damage, Sustained DNA Double-Strand Breaks, and the ATM/ATR Damage Response in Cancer Cells. *Mol. Cancer Ther.* **2011**, *10*, 2224–2233. [CrossRef] [PubMed]
82. Efferth, T.; Dunstan, H.; Sauerbrey, A.; Miyachi, H.; Chitambar, C. The anti-malarial artesunate is also active against cancer. *Int. J. Oncol.* **2001**, *18*, 767–773. [CrossRef] [PubMed]
83. Bakhoum, S.F.; Kabeche, L.; Compton, D.A.; Powell, S.N.; Bastians, H. Mitotic DNA Damage Response: At the Crossroads of Structural and Numerical Cancer Chromosome Instabilities. *Trends Cancer* **2017**, *3*, 225–234. [CrossRef] [PubMed]
84. Nogueira, V.; Hay, N. Molecular pathways: Reactive oxygen species homeostasis in cancer cells and implications for cancer therapy. *Clin. Cancer Res.* **2013**, *19*, 4309–4314. [CrossRef] [PubMed]

cells

MDPI

Review

Regulation of Autophagy Is a Novel Tumorigenesis-Related Activity of Multifunctional Translationally Controlled Tumor Protein

Ji-Sun Lee, Eun-Hwa Jang, Hyun Ae Woo and Kyunglim Lee *

Graduate School of Pharmaceutical Sciences, College of Pharmacy, Ewha Womans University, Seoul 03760, Korea; leejisun78@hanmail.net (J.-S.L.); ehj_@naver.com (E.-H.J.); hawoo@ewha.ac.kr (H.A.W.)

* Correspondence: klyoon@ewha.ac.kr; Tel.: +82-2-3277-3024

Received: 14 November 2019; Accepted: 17 January 2020; Published: 20 January 2020

Abstract: Translationally controlled tumor protein (TCTP) is highly conserved in eukaryotic organisms and plays multiple roles regulating cellular growth and homeostasis. Because of its anti-apoptotic activity and its role in the regulation of cancer metastasis, TCTP has become a promising target for cancer therapy. Moreover, growing evidence points to its clinical role in cancer prognosis. How TCTP regulates cellular growth in cancer has been widely studied, but how it regulates cellular homeostasis has received relatively little attention. This review discusses how TCTP is related to cancer and its potential as a target in cancer therapeutics, including its novel role in the regulation of autophagy. Regulation of autophagy is essential for cell recycling and scavenging cellular materials to sustain cell survival under the metabolic stress that cancer cells undergo during their aggressive proliferation.

Keywords: TCTP; cancer; autophagy

1. Role of TCTP in Tumorigenesis

TCTP (also known as histamine releasing factor, HRF; fortilin) is a highly conserved and multifunctional protein that participates in diverse biological and disease processes, including cancer [1–3]. Many studies have established the diverse roles of TCTP in tumorigenesis. The best characterized function of TCTP is its anti-apoptotic activity [2,4–6]. The anti-apoptotic role of TCTP was first identified in a study that showed the overexpression of TCTP in various cancer cells prevented etoposide-mediated apoptosis [7]. TCTP was shown to exert its anti-apoptotic activity by enhancing the stability of Mcl-1, a member of Bcl-2 family proteins [8]. Further, it was shown that depletion of TCTP in mice led to increased apoptosis during embryogenesis and that lethality supported its role in mediating apoptosis [9]. Crystallography studies revealed that the helical domain of TCTP displays similarity to helices H5-H6 of Bax which have been known to regulate mitochondrial membrane permeability during apoptosis. Based on this finding, a novel mechanism for the antiapoptotic activity of TCTP, which involves its localization to the mitochondria and inhibition of Bax dimerization was proposed [9]. TCTP also potentiates the antiapoptotic function of Bcl-xL through its BH3-like domain [10]. Moreover, genomic integrity is maintained by TCTP via its interaction with ataxia-telangiectasia mutated (ATM) and p53 [11,12]. Low-dose γ-irradiation induced localization of TCTP into the nucleus where it exists complexed with other proteins that participate in DNA damage sensing and repair [11].

TCTP is also implicated in metastasis, tumor invasion, and resistance to anticancer therapy [13–16]. Knockdown of TCTP in colon adenocarcinoma inhibited proliferation, migration and invasion of tumor cells both in vitro and in vivo [17]. Proteomic analysis revealed that the expression of proteins related to cytoskeleton biosynthesis was changed by TCTP knockdown [13]. In colorectal cancer, extracellular TCTP promoted disease progression and liver metastasis through Cdc42/JNK/MMP9 activation [14]. It was also suggested that TCTP regulates metastasis of colorectal cancer by regulation of the high mobility group 1

(HMGB1) and activation of the NF-KB pathway [17], or by regulating epithelial-mesenchymal transition (EMT) via mammalian target of the rapamycin complex 2 (mTORC2)/Akt/GSK3β/β-catenin pathway [15].

High expression of TCTP in a variety of cancers, including cholangiocarcinoma [18], ovarian cancer [19], and glioma [20], correlates with poor prognosis. Immunohistochemical analysis of normal and colorectal cancer tissues revealed that TCTP expression is significantly higher in the tumor tissues. Elevated TCTP was observed from the adenoma stage, indicating that TCTP is induced early in the disease, and further induction in later stages was not observed [21]. In a cohort of breast cancer patients, TCTP levels increased in later stages of cancer with the poor prognosis [22]. A significant association of TCTP in advanced stages of human hepatocellular carcinoma was observed. This study revealed transcriptional regulation mediated by the chromodomain helicase/ATPase DNA binding protein 1-like gene (CHD1L) [23].

TCTP plays multiple roles in the pathogenesis of cancer, including regulation of resistance to cancer therapy, especially including chemotherapy and radiation therapy [16,21,24]. Jung et al. showed that overexpression of TCTP in HeLa cells inhibited cell death by cytotoxic drugs through the inhibition of mitochondria-mediated apoptosis [16]. During the apoptosome formation induced by chemotherapeutic agents, C-terminally cleaved TCTP associates with Apaf-1 in the apoptosome, thus preventing the caspase activation cascade. In colorectal cancer cells, 5-FU and oxaliplatin treatment induced translational upregulation of TCTP via the mTORC1 pathway and decreased TCTP expression increased the sensitivity to the chemotherapeutic agents [21]. Moreover, breast and lung cancer cells with relatively high expression levels of TCTP were insensitive to radiation-induced cell death and response to radiation was influenced by TCTP levels [24].

Because of its anti-apoptotic properties, its role in the regulation of cancer metastasis, and its relevance to disease prognosis, TCTP has become a promising target for cancer therapy (Table 1). Mechanisms underlying how TCTP regulates cellular growth in cancer have been widely studied, but the question whether autophagy may be involved in how TCTP regulates cellular homeostasis has received relatively little attention. In this review, we discuss TCTP's role in cancer pathogenesis, especially in relation to its novel role in the regulation of autophagy and its potential as a target in cancer therapeutics.

Table 1. Roles of translationally controlled tumor protein (TCTP) in cancer pathogenesis.

Function	Mechanism	Reference
Anti-apoptosis	Increased Mcl-1 stability	[8]
	Inhibition of Bax dimerization	[9]
	Interaction with ATM and p53	[11,12]
	Promotion of MDM2-mediate p53 degradation	[22]
Pro-metastasis	Cdc42/JNK/MMP9 activation	[14]
	Regulation of HMGB1/NF-kB activation	[17]
	Promotion of EMT through mTORC2/Akt/GSK3β/β-catenin pathway	[15]
Resistance to anti-cancer therapy	Prevention of apoptosis through association with Apaf-1	[16]
	Translational upregulation of TCTP via mTORC1 pathway	[21]
	Promotion of MDM2-mediated p53 degradation	[24]

2. Autophagy, a Machinery for Cellular Homeostasis

Autophagy is the process that delivers cellular proteins and organelles to lysosomes for digestion by lysosomal hydrolases, thereby facilitating metabolic homeostasis [25–29]. By eliminating redundant proteins and damaged organelles and by providing building blocks and energy, autophagy serves as a recycling system for cell renewal. During metabolic stress, autophagy prevents the accumulation of toxic cellular components, enabling cells to adapt to environmental changes [29]. Autophagy occurs both in physiological stresses such as nutrient deprivation, ER stress, and hypoxia, as well as in pathological states, including infection, cancer, and neurodegeneration. Therefore, the modulation of autophagy as a therapeutic strategy has recently received increasing attention. But whether autophagy is the cause or result of disease is still an unsettled question. While it is an important catabolic survival mechanism, autophagy has also been classified as type II programmed cell death based on the observation that the cytoplasm of dying cells contains plenty of autophagic vacuolization [26]. However, whether autophagy results in cell death or merely accompanies cell death is still debated [30].

During the autophagic process, cells form double-membrane vesicles called autophagosomes. Lysosome then fuses with the autophagosomes to form autolysosomes. This results in the degradation of the cellular components originally captured by the autophagosomes and in the recycling of their building blocks. This process is divided into five steps: initiation, vesicle nucleation, vesicle elongation, vesicle fusion, and cargo degradation [31]. When cells are in a nutrient-rich environment, the mTORC1 complex inhibits autophagy by preventing the formation of the ULK1 complex (Figure 1) [32,33]. In the initiation step, the ULK1 complex dissociates from mTORC1 and activates the class III PI3K complex. In this complex, Beclin 1 works as a scaffold protein to recruit VPS34, ATG15, UV radiation resistance-associated gene protein (UVRAG), and an activating molecule in BECN1-regulated autophagy protein 1 (AMBRA1). The class III PI3K complex is the main mediator of vesicle nucleation by localizing autophagic proteins into phagophores [33,34].

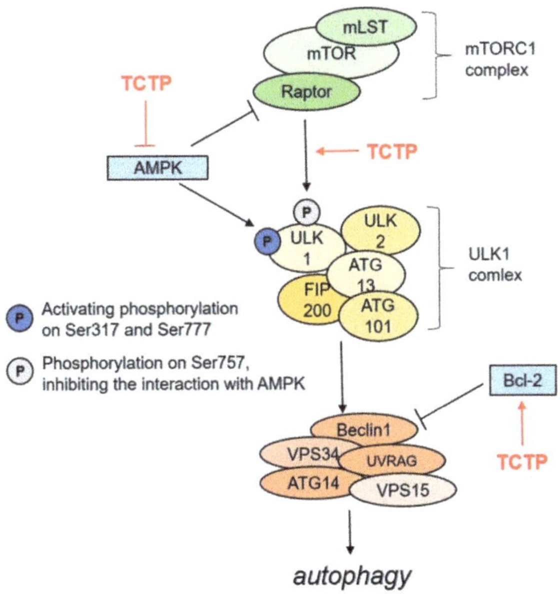

Figure 1. Negative regulation of autophagy by TCTP. In the initiation step of autophagy, TCTP may inhibit AMPK-induced ULK1 phosphorylation on Ser317 and Ser777, which activates the formation of the Beclin 1 complex. TCTP also enhances the mTORC1-induced phosphorylation of ULK1 on Ser757, which prevents ULK1 interaction with AMPK. TCTP also inhibits autophagosome formation by activating Bcl-2, which, in turn, inhibits the formation of the Beclin1 complex (see text for further details).

Once the phagophore, a double-membraned opened structure, has captured its cargo material, the autophagosome undergoes maturation by vesicle elongation process, which is regulated by two ubiquitin-like conjugation systems. The first system involves the incorporation of LC3II into a growing double-membrane, and it is regulated by the protease ATG4B and by the E1-like enzyme ATG7 [35].

The other system requires the formation of the ATG12-ATG5·ATG16L1 complex, which is mediated by ATG7 and by the E2-like enzyme ATG10 [36,37]. In the vesicle fusion step, the autophagosome fuses with the lysosome, facilitated by syntaxin 17 (STX17) [38], and the autophagosome contents are degraded due to the action of the lysosomal enzymes. Autophagy plays an indispensable role in these cellular metabolic events because absence or defective autophagy leads to metabolic, neurodegenerative, and infectious diseases [39]. Defective autophagy has also been reported to cause tumorigenesis, facilitated by several protein partners including TCTP. TCTP has been shown to be involved tumorigenesis, but its specific role in the regulation of autophagy, of which little is known, is discussed in Section 6 below.

3. Role of Autophagy in Tumorigenesis

Autophagy plays a dual role in regulating tumorigenesis: (A) suppressing tumor growth and (B) maintaining the survival of cancer cells. Suppression of tumor growth by inhibiting the proliferation of cancer cells was initially considered as the primary role of autophagy [40]. Exogenous expression of Beclin 1 in human breast cancer carcinoma cells induced autophagy and this was associated with decreased cell proliferation and tumorigenesis in nude mice [41]. Studies in a mutant mouse model revealed BECN1 as a haploinsufficient tumor suppressor gene [42] and provided evidence for negative regulation of tumorigenesis by Beclin-1 [43]. As a possible mechanism for tumor suppression by autophagy, it was proposed that autophagy prevents the accumulation of oxidative stress which would result in genomic instability [27]. During autophagy deficiency, accumulated p62, also known as sequestosome 1 (SQSTM-1), prevents Keap1-mediated ubiquitination of Nrf2, thereby activating an antioxidant defense pathway [44,45]. Another proposed mechanism suggests that autophagy suppresses tumorigenesis through inhibition of inflammation. When autophagy is deficient in cancer cells, necrotic cell death induces an inflammatory response and secretion of pro-inflammatory factors such as the HMGB1 protein, eventually leading to tumor growth [25]. Under chronic inflammation, a proliferation of cells occurs to compensate for the damage from inflammation-inducing cells that are susceptible to genetic mutations and oncogenic transformation [46]. Thus autophagy suppresses inflammation-induced tumorigenesis by mitigating cellular damage [47]. Autophagy also suppresses the proliferation of oncogene-expressing cancer cells by inducing senescence, which functions as a barrier to tumorigenesis [48]. Liu et al. showed that downregulation of ATG5 expression enhanced proliferation and prevented oncogene-induced senescence in melanocytes, thus confirming ATG5 as tumor suppressor gene [49]. Autophagic activity is also upregulated during oncogene-induced senescence and overexpression of ULK3, an autophagy-related gene, resulted in senescence [50]. Thus, autophagy was postulated to suppress tumorigenesis by forcing oncogene-expressing cells to exit the cell cycle.

Cancer cells with high metabolic demand due to increased proliferation depend on autophagy to maintain cancer cell survival [29]. Several studies have reported that autophagy supports cancer cell survival, proliferation, and metastasis-related behaviors. Ectopic expression of H-ras^{v12} or K-ras^{v12} in nontumorigenic epithelial cells showed increased basal autophagy, which supports the growth of RAS-driven tumors by maintaining functional mitochondria under the nutrient-limited condition [51]. Moreover, when autophagy is blocked by knockout of ATG5 or ATG7 in these cells, tumor growth in vivo is delayed [51]. Studies with a breast cancer mouse model also demonstrated the tumor-promoting role of autophagy [52]. In this study, conditional knockout of the protein, FIP200, essential for autophagy, inhibited breast cancer initiation, progression, and metastasis. Gene expression profiles of FIP200-knocked out mammary tumors revealed increased expression of immune response genes, suggesting that FIP200 deficiency induces anti-tumor immune surveillance that eventually suppresses tumor progression [52]. Moreover, autophagy inhibition by hydroxychloroquine (HCQ) decreased cellular proliferation and metastasis of dormant breast cancer cells. In this study, knockdown of ATG7 reduced the metastatic burden, while knockdown of BECN1 showed no effects, suggesting that dormant breast cancer cells are autophagic, which is dependent on ATG7 [44]. Additionally, autophagy serves as a survival mechanism for cells under tumor-related conditions, such as hypoxia or metabolic stress. Under hypoxic conditions, hypoxia-inducible factor-1α (HIF-1α) induces mitochondrial hypoxia as an adaptive response to metabolic stress, preventing cell death [53]. With regard to cellular adaptation, autophagy provides cancer cells with a strategy to survive under metabolic stress conditions with limited nutrient and oxygen supply.

Overall, the effect of autophagy on tumorigenesis depends on cancer type, stage, and cancer environment [28,54]. Since genomic instability can initiate cancer [55], autophagy can delay tumorigenesis by maintaining genomic integrity. However, once tumors are formed and begin to proliferate, cancer cells themselves take advantage of autophagy to sustain their rapid proliferation.

4. Autophagy, a Potential Target for Cancer Treatment

Probably due to the dual role of autophagy in cancer, both as an inhibitor and as an activator, there is an increasing interest to investigate the modulation of autophagy as a potential avenue for cancer therapy. The application of pharmacological modulators of autophagy has shown promise in some in vitro and in vivo studies, but not in others [56–58]. Several studies have reported that treatment with anti-cancer drugs resulted in the induction of autophagy in a range of cancer cell lines [59]. In human mammary carcinoma cells, tamoxifen treatment resulted in autophagic vacuole formation, and 3-methyladenine (3-MA), an inhibitor of autophagosome vacuoles, partially inhibited cell death caused by tamoxifen [60], suggesting that autophagy synergized with apoptosis to cause cell death under the tamoxifen treatment. However, more recent findings suggest that autophagy serves as a possible mechanism of tamoxifen resistance in breast cancer cells [61–63]. These studies imply that autophagy induced by anticancer therapy may serve as a cellular protection phenomenon.

Besides, autophagy is also related to radiation-induced cancer cell death. Human breast cancer cells, MCF7, are sensitive to standard doses of radiation. However, the accumulation of acidic vesicular organelles, a characteristic feature of autophagy, was observed in surviving cell population after the irradiation and a potent autophagy inhibitor bafilomycin A enhanced apoptosis-related cell death after the irradiation [64]. These findings suggest that autophagy induced by conventional cancer therapy could be a potential mechanism for the development of therapy resistance by providing cells with an alternative survival pathway to avoid apoptotic cell death. Thus, research on autophagy in the context of cancer therapy and considering autophagy inhibitors in combination with conventional cancer therapy could be a strategy to overcome therapy resistance.

Agents that inhibit autophagy have been suggested as anti-cancer agents, and most class III PI3K inhibitors are categorized to this type. Class III PI3K inhibitors mediate autophagosome formation, and agents such as 3-MA, wortmannin, and LY294002 have been shown to inhibit autophagy [65]. 3-MA, in particular, is distinguished from other PI3K inhibitors in that it exerts a dual role in autophagy. In nutrient-rich conditions, 3-MA promoted autophagy flux while still showing inhibitory effects on starvation-induced autophagy [66]. These differential effects on autophagy arise from the divergent effects of 3-MA on PI3K subunits. While blocking by 3-MA of class I PI3K is persistent, its inhibition of class III PI3K is transient. Under nutrient-rich conditions, prolonged treatment with 3-MA continuously blocks class I PI3K and the Akt-mTOR pathway, which inhibits autophagy while sparing class III PI3K, eventually activating autophagy [66]. Since class I PI3K has been known to be a negative regulator of autophagy [67], selecting PI3K inhibitors as autophagy-regulating anti-cancer agents should take into consideration of how they affect each subtype of PI3K.

Also, lysosomotropic agents such as chloroquine (CQ), HCQ, Lys0569, or monensin have been recommended for curing cancer by preventing acidification of lysosome and eventually inhibiting degradation of molecules in autophagosome [68,69]. Indeed CQ and HCQ are registered anti-malarial drugs and have shown anticancer effects when administered in combination with conventional chemotherapy and radiotherapy [70–72]. Furthermore, Bafilomycin A interferes with both lysosome acidification and autophagosome-lysosome fusion and this observation may have the potential to lead to the development of anti-cancer agents that overcome resistance to anticancer therapy [73,74].

A potent autophagy inhibitor spautin-1 [75] also has been reported to enhance the potency of imatinib in chronic myeloid leukemia [76].

On the other hand, inducing excessive autophagy is also suggested for treating cancer, especially in cancer cells that are resistant to apoptotic stimuli. The most commonly studied agent for inducing autophagy is rapamycin, which inhibits mTOR, thus activating autophagy by preventing the inactivating phosphorylation of ULK1. Although rapamycin is an FDA-approved drug for cancers, such as renal cancer, its clinical use has limitations because different cancer cells respond differently to rapamycin [77]. Tyrosine kinase inhibitors and histone deacetylase inhibitors are also known to induce autophagy in cancer cells, but further investigations are needed to settle whether autophagy is induced by cells that are resistant to those inhibitors or as a readout of cell death. It is important to establish whether the induced autophagy results in drug resistance or cell death. Therefore, mechanistic studies about current cancer therapy and autophagy are still worthwhile for enhancing therapeutic efficacy, although autophagy plays a complex role in tumorigenesis.

5. Autophagy-Mediated TCTP Degradation

The relationship between TCTP and autophagy has not received much investigative interest, and only a few studies have reported the regulation by and effect on TCTP by autophagy. The expression of TCTP is fine-tuned both by transcriptional and translational regulation by a variety of extracellular signals, but its degradation mechanism is poorly understood. Several studies reported that interaction with several proteins, including Hsp27 and Mcl-1, increase the protein stability of TCTP [78,79] and those studies suggested ubiquitin-proteasome mediated degradation as a TCTP protein regulation mechanism. However, recently, a new type of regulation called chaperone-mediated autophagy (CMA) was suggested as a possible mechanism for TCTP degradation [80]. Unlike in macroautophagy where the substrates are sequestered to the vesicles, which then fuse with lysosomes, in CMA, proteins are directly trapped inside vesicles by the invagination of the lysosomal membrane. Only proteins targeted to the lysosome can be taken up into the lumen of the lysosome [81]. Thus CMA works as a unique proteolytic system that degrades specific cytosolic proteins, which permits quality control of proteins. Thus regulation of CMA could be a promising strategy to combat cancer, as shown by several studies [82–84]. In mouse embryonic fibroblasts (MEFs), serum starvation decreased cellular TCTP expression and this was found to be independent of macroautophagy and the ubiquitin-proteasome pathway but was dependent on the 70 kDa heat shock cognate protein (Hsc70) and the lysosome-associated membrane protein type 2A (LAMP-2A) [80]. Hsc70 is a cytosolic chaperone that recognizes the KFERQ motif in the amino acid sequence of the substrate protein, which is mandatory for targeting to the lysosome membrane [85]. LAMP-2A works as a lysosomal receptor for the substrate protein–chaperone complex [86]. Moreover, the KFERQ-like motif in TCTP was generated by acetylation of Lys19, and this increased the binding of TCTP to Hsc70 [80].

6. The role of TCTP in the Regulation of Autophagy

Since TCTP influences cell survival and proliferation, it has the potential to influence autophagy, most likely through its association with the mTORC1 signaling pathway. Being at the center of a signaling network, the mTORC1 pathway regulates many processes, which are important for cell growth. This pathway also negatively regulates autophagy by competing with adenosine monophosphate-activated protein kinase (AMPK). During glucose starvation, AMPK drives autophagy by phosphorylation of Ser317 and Ser777 in autophagy initiator, ULK1. When sufficient glucose is provided, mTORC1 phosphorylates Ser757 of ULK1, which prevents the interaction between ULK1 and AMPK, eventually blocking autophagy [87]. We have reported that TCTP is able to interfere with these processes at several points [88], and thereby negatively regulates autophagy (Figure 1). TCTP is translationally induced by growth stimulation through the mTORC1 pathway [89], and therefore regulation of cell growth by TCTP may be related to autophagy. Indeed, decreased TCTP expression promoted formation and maturation of autophagosomes, observed by LC3 puncta formation and by co-localization of LC3 with the lysosomal marker, LAMP1 [88]. Further studies revealed that downregulation of TCTP in HeLa cells resulted in reduced levels of downstream effectors of the mTORC1 pathway, including p-EIF-4EBP1, p-RPS6KB, p-ULK1 (Ser757), but increased AMPK phosphorylation [88]. Interestingly, TCTP knockdown decreased the mTORC1 pathway synergistically with rapamycin, a well-known allosteric inhibitor of mTOR [90], suggesting that TCTP might be a potential target to antagonize rapamycin resistance in cancer therapy. TCTP also regulates autophagy in an AMPK-mTORC1-pathway-independent mode by modulating the formation of the Beclin-1 complex [88]. Beclin-1 is a mammalian orthologue of yeast ATG-6 [91] and performs a central role as overall scaffold for the class III PI3K complex [92]. Additionally, Bcl-2 also blocks autophagy by binding to Beclin-1, disrupting the Beclin-1/PI3K complex [93]. TCTP knockdown reduced bcl-2 expression without affecting Beclin-1 expression and thereby promoted Beclin-1-mediated autophagosome formation [88]. Negatively regulated autophagy by TCTP was further confirmed by a mouse model, in which TCTP is haploinsufficiently expressed [88]. Although more detailed mechanistic studies need to be conducted, we can conclude from the present study that TCTP, because it is involved in regulating cellular autophagy, could be a potential target in cancer therapy.

7. Conclusions

TCTP participates in a wide variety of cancer-related phenomena including cancer progression, regulation of apoptosis, tumor reversion, and development of resistance to anti-cancer therapy. This suggests that pharmacological interventions that inhibit TCTP's functions are rational targets to be considered in cancer therapy. Regrettably, no attempts in this regard have been reported yet. Because of the diverse biological functions of TCTP such as development, growth, protein synthesis, and allergic response, targeting TCTP toward cancer cure should avoid systemic inhibition of this molecule and the potential risks to normal cells. A recent study of TCTP-induced negative regulation of autophagy has provided another mechanism for TCTP regulation of cancer progression. Since autophagy is the main machinery for the regulation of cellular homeostasis, TCTP also has the potential to affect cancer cell metabolism, which is different from conventional roles of TCTP in cancer, such as regulation of cancer cell proliferation. However, the consequences of targeting TCTP in cancer should be more precisely investigated because the role of autophagy in cancer is also dependent on the type and pathological stages of the specific tumor. In the early stages of tumorigenesis, which include initiation of malignant transformation and progression, autophagy may contribute to suppressing tumor development. However, in later stages, autophagy may help to sustain tumor growth by supporting energy production and by providing building blocks for cellular syntheses. Further mechanistic understanding of how TCTP mediates autophagy in cancer should help answer the question of whether TCTP can serve as a target in cancer therapy.

Author Contributions: K.L. and H.A.W. conceived and designed the review. J.-S.L. mainly wrote the review and E.-H.J. added revisions as necessary. All authors have read and agreed to the published version of the manuscript.

Funding: This work is partially supported by Basic Science Research Program (2017R1A2B2004023) and Bio & Medical Technology Development Program (2018M3A9A8021689) through the National Research Foundation of Korea (NRF) funded by the Ministry of Science, ICT and future Planning.

Conflicts of Interest: The authors declare no conflict of interest.

References

1. Jung, J.; Ryu, S.; Ki, I.A.; Woo, H.A.; Lee, K. Some Biological Consequences of the Inhibition of Na,K-ATPase by Translationally Controlled Tumor Protein (TCTP). *Int. J. Mol. Sci.* **2018**, *19*, 1657. [CrossRef] [PubMed]
2. Acunzo, J.; Baylot, V.; So, A.; Rocchi, P. TCTP as therapeutic target in cancers. *Cancer Treat. Rev.* **2014**, *40*, 760–769. [CrossRef] [PubMed]
3. Bommer, U.A.; Thiele, B.J. The translationally controlled tumour protein (TCTP). *Int. J. Biochem. Cell Biol.* **2004**, *36*, 379–385. [CrossRef]

4. Koziol, M.J.; Gurdon, J.B. TCTP in development and cancer. *Biochem. Res. Int.* **2012**, *2012*, 105203. [CrossRef] [PubMed]
5. Gnanasekar, M.; Thirugnanam, S.; Zheng, G.; Chen, A.; Ramaswamy, K. Gene silencing of translationally controlled tumor protein (TCTP) by siRNA inhibits cell growth and induces apoptosis of human prostate cancer cells. *Int. J. Oncol.* **2009**, *34*, 1241–1246. [CrossRef] [PubMed]
6. Lucibello, M.; Gambacurta, A.; Zonfrillo, M.; Pierimarchi, P.; Serafino, A.; Rasi, G.; Rubartelli, A.; Garaci, E. TCTP is a critical survival factor that protects cancer cells from oxidative stress-induced cell-death. *Exp. Cell Res.* **2011**, *317*, 2479–2489. [CrossRef]
7. Li, F.; Zhang, D.; Fujise, K. Characterization of fortilin, a novel antiapoptotic protein. *J. Biol. Chem.* **2001**, *276*, 47542–47549. [CrossRef]
8. Liu, H.; Peng, H.W.; Cheng, Y.S.; Yuan, H.S.; Yang-Yen, H.F. Stabilization and enhancement of the antiapoptotic activity of mcl-1 by TCTP. *Mol. Cell Biol.* **2005**, *25*, 3117–3126. [CrossRef]
9. Susini, L.; Besse, S.; Duflaut, D.; Lespagnol, A.; Beekman, C.; Fiucci, G.; Atkinson, A.R.; Busso, D.; Poussin, P.; Marine, J.C.; et al. TCTP protects from apoptotic cell death by antagonizing bax function. *Cell Death Differ.* **2008**, *15*, 1211–1220. [CrossRef]
10. Thebault, S.; Agez, M.; Chi, X.; Stojko, J.; Cura, V.; Telerman, S.B.; Maillet, L.; Gautier, F.; Billas-Massobrio, I.; Birck, C.; et al. TCTP contains a BH3-like domain, which instead of inhibiting, activates Bcl-xL. *Sci. Rep.* **2016**, *6*, 19725. [CrossRef]
11. Zhang, J.; de Toledo, S.M.; Pandey, B.N.; Guo, G.; Pain, D.; Li, H.; Azzam, E.I. Role of the translationally controlled tumor protein in DNA damage sensing and repair. *Proc. Natl. Acad. Sci. USA* **2012**, *109*, E926–E933. [CrossRef] [PubMed]
12. Hong, S.T.; Choi, K.W. TCTP directly regulates ATM activity to control genome stability and organ development in Drosophila melanogaster. *Nat. Commun.* **2013**, *4*, 2986. [CrossRef] [PubMed]
13. Ma, Q.; Geng, Y.; Xu, W.; Wu, Y.; He, F.; Shu, W.; Huang, M.; Du, H.; Li, M. The role of translationally controlled tumor protein in tumor growth and metastasis of colon adenocarcinoma cells. *J. Proteome Res.* **2010**, *9*, 40–49. [CrossRef] [PubMed]
14. Xiao, B.; Chen, D.; Luo, S.; Hao, W.; Jing, F.; Liu, T.; Wang, S.; Geng, Y.; Li, L.; Xu, W.; et al. Extracellular translationally controlled tumor protein promotes colorectal cancer invasion and metastasis through Cdc42/JNK/MMP9 signaling. *Oncotarget* **2016**, *7*, 50057–50073. [CrossRef] [PubMed]
15. Bae, S.Y.; Kim, H.J.; Lee, K.J.; Lee, K. Translationally controlled tumor protein induces epithelial to mesenchymal transition and promotes cell migration, invasion and metastasis. *Sci. Rep.* **2015**, *5*, 8061. [CrossRef] [PubMed]
16. Jung, J.; Kim, H.Y.; Maeng, J.; Kim, M.; Shin, D.H.; Lee, K. Interaction of translationally controlled tumor protein with Apaf-1 is involved in the development of chemoresistance in HeLa cells. *BMC Cancer* **2014**, *14*, 165. [CrossRef]
17. Huang, M.; Geng, Y.; Deng, Q.; Li, R.; Shao, X.; Zhang, Z.; Xu, W.; Wu, Y.; Ma, Q. Translationally controlled tumor protein affects colorectal cancer metastasis through the high mobility group box 1-dependent pathway. *Int. J. Oncol.* **2018**, *53*, 1481–1492. [CrossRef]
18. Phanthaphol, N.; Techasen, A.; Loilome, W.; Thongchot, S.; Thanan, R.; Sungkhamanon, S.; Khuntikeo, N.; Yongvanit, P.; Namwat, N. Upregulation of TCTP is associated with cholangiocarcinoma progression and metastasis. *Oncol Lett* **2017**, *14*, 5973–5979. [CrossRef]
19. Chen, C.; Deng, Y.; Hua, M.; Xi, Q.; Liu, R.; Yang, S.; Liu, J.; Zhong, J.; Tang, M.; Lu, S.; et al. Expression and clinical role of TCTP in epithelial ovarian cancer. *J. Mol. Histol.* **2015**, *46*, 145–156. [CrossRef]
20. Miao, X.; Chen, Y.B.; Xu, S.L.; Zhao, T.; Liu, J.Y.; Li, Y.R.; Wang, J.; Zhang, J.; Guo, G.Z. TCTP overexpression is associated with the development and progression of glioma. *Tumour. Biol.* **2013**, *34*, 3357–3361. [CrossRef]
21. Bommer, U.A.; Vine, K.L.; Puri, P.; Engel, M.; Belfiore, L.; Fildes, K.; Batterham, M.; Lochhead, A.; Aghmesheh, M. Translationally controlled tumour protein TCTP is induced early in human colorectal tumours and contributes to the resistance of HCT116 colon cancer cells to 5-FU and oxaliplatin. *Cell Commun. Signal.* **2017**, *15*, 9. [CrossRef] [PubMed]
22. Amson, R.; Pece, S.; Lespagnol, A.; Vyas, R.; Mazzarol, G.; Tosoni, D.; Colaluca, I.; Viale, G.; Rodrigues-Ferreira, S.; Wynendaele, J.; et al. Reciprocal repression between P53 and TCTP. *Nat. Med.* **2011**, *18*, 91–99. [CrossRef] [PubMed]

23. Chan, T.H.; Chen, L.; Liu, M.; Hu, L.; Zheng, B.J.; Poon, V.K.; Huang, P.; Yuan, Y.F.; Huang, J.D.; Yang, J.; et al. Translationally controlled tumor protein induces mitotic defects and chromosome missegregation in hepatocellular carcinoma development. *Hepatology* **2012**, *55*, 491–505. [CrossRef] [PubMed]
24. Jung, J.; Lee, J.S.; Lee, Y.S.; Lee, K. Radiosensitivity of Cancer Cells Is Regulated by Translationally Controlled Tumor Protein. *Cancers* **2019**, *11*, 386. [CrossRef] [PubMed]
25. Degenhardt, K.; Mathew, R.; Beaudoin, B.; Bray, K.; Anderson, D.; Chen, G.; Mukherjee, C.; Shi, Y.; Gelinas, C.; Fan, Y.; et al. Autophagy promotes tumor cell survival and restricts necrosis, inflammation, and tumorigenesis. *Cancer Cell* **2006**, *10*, 51–64. [CrossRef] [PubMed]
26. Kondo, Y.; Kondo, S. Autophagy and cancer therapy. *Autophagy* **2006**, *2*, 85–90. [CrossRef] [PubMed]
27. Chen, H.Y.; White, E. Role of autophagy in cancer prevention. *Cancer Prev Res. (Phila)* **2011**, *4*, 973–983. [CrossRef]
28. Kimmelman, A.C. The dynamic nature of autophagy in cancer. *Genes Dev.* **2011**, *25*, 1999–2010. [CrossRef]
29. White, E. The role for autophagy in cancer. *J. Clin. Invest.* **2015**, *125*, 42–46. [CrossRef]
30. Kroemer, G.; Levine, B. Autophagic cell death: The story of a misnomer. *Nat. Rev. Mol. Cell Biol.* **2008**, *9*, 1004–1010. [CrossRef]
31. Levy, J.M.M.; Towers, C.G.; Thorburn, A. Targeting autophagy in cancer. *Nat. Rev. Cancer* **2017**, *17*, 528–542. [CrossRef] [PubMed]
32. Papinski, D.; Kraft, C. Regulation of Autophagy By Signaling Through the Atg1/ULK1 Complex. *J. Mol. Biol.* **2016**, *428*, 1725–1741. [CrossRef] [PubMed]
33. Russell, R.C.; Tian, Y.; Yuan, H.; Park, H.W.; Chang, Y.Y.; Kim, J.; Kim, H.; Neufeld, T.P.; Dillin, A.; Guan, K.L. ULK1 induces autophagy by phosphorylating Beclin-1 and activating VPS34 lipid kinase. *Nat. Cell Biol.* **2013**, *15*, 741–750. [CrossRef] [PubMed]
34. Funderburk, S.F.; Wang, Q.J.; Yue, Z. The Beclin 1-VPS34 complex—At the crossroads of autophagy and beyond. *Trends Cell Biol.* **2010**, *20*, 355–362. [CrossRef]
35. Mizushima, N.; Noda, T.; Yoshimori, T.; Tanaka, Y.; Ishii, T.; George, M.D.; Klionsky, D.J.; Ohsumi, M.; Ohsumi, Y. A protein conjugation system essential for autophagy. *Nature* **1998**, *395*, 395–398. [CrossRef]
36. Shintani, T.; Mizushima, N.; Ogawa, Y.; Matsuura, A.; Noda, T.; Ohsumi, Y. Apg10p, a novel protein-conjugating enzyme essential for autophagy in yeast. *EMBO J.* **1999**, *18*, 5234–5241. [CrossRef]
37. Mizushima, N.; Kuma, A.; Kobayashi, Y.; Yamamoto, A.; Matsubae, M.; Takao, T.; Natsume, T.; Ohsumi, Y.; Yoshimori, T. Mouse Apg16L, a novel WD-repeat protein, targets to the autophagic isolation membrane with the Apg12-Apg5 conjugate. *J. Cell Sci.* **2003**, *116*, 1679–1688. [CrossRef]
38. Itakura, E.; Kishi-Itakura, C.; Mizushima, N. The hairpin-type tail-anchored SNARE syntaxin 17 targets to autophagosomes for fusion with endosomes/lysosomes. *Cell* **2012**, *151*, 1256–1269. [CrossRef]
39. Jiang, P.; Mizushima, N. Autophagy and human diseases. *Cell Res.* **2014**, *24*, 69–79. [CrossRef]
40. Mathew, R.; Karp, C.M.; Beaudoin, B.; Vuong, N.; Chen, G.; Chen, H.Y.; Bray, K.; Reddy, A.; Bhanot, G.; Gelinas, C.; et al. Autophagy suppresses tumorigenesis through elimination of p62. *Cell* **2009**, *137*, 1062–1075. [CrossRef]
41. Liang, X.H.; Jackson, S.; Seaman, M.; Brown, K.; Kempkes, B.; Hibshoosh, H.; Levine, B. Induction of autophagy and inhibition of tumorigenesis by beclin 1. *Nature* **1999**, *402*, 672–676. [CrossRef] [PubMed]
42. Yue, Z.; Jin, S.; Yang, C.; Levine, A.J.; Heintz, N. Beclin 1, an autophagy gene essential for early embryonic development, is a haploinsufficient tumor suppressor. *Proc. Natl. Acad. Sci. USA* **2003**, *100*, 15077–15082. [CrossRef] [PubMed]
43. Qu, X.; Yu, J.; Bhagat, G.; Furuya, N.; Hibshoosh, H.; Troxel, A.; Rosen, J.; Eskelinen, E.L.; Mizushima, N.; Ohsumi, Y.; et al. Promotion of tumorigenesis by heterozygous disruption of the beclin 1 autophagy gene. *J. Clin. Invest.* **2003**, *112*, 1809–1820. [CrossRef] [PubMed]
44. Lau, A.; Wang, X.J.; Zhao, F.; Villeneuve, N.F.; Wu, T.; Jiang, T.; Sun, Z.; White, E.; Zhang, D.D. A noncanonical mechanism of Nrf2 activation by autophagy deficiency: Direct interaction between Keap1 and p62. *Mol. Cell Biol.* **2010**, *30*, 3275–3285. [CrossRef]
45. Komatsu, M.; Kurokawa, H.; Waguri, S.; Taguchi, K.; Kobayashi, A.; Ichimura, Y.; Sou, Y.S.; Ueno, I.; Sakamoto, A.; Tong, K.I.; et al. The selective autophagy substrate p62 activates the stress responsive transcription factor Nrf2 through inactivation of Keap1. *Nat. Cell Biol.* **2010**, *12*, 213–223. [CrossRef]
46. Grivennikov, S.I.; Greten, F.R.; Karin, M. Immunity, inflammation, and cancer. *Cell* **2010**, *140*, 883–899. [CrossRef]

47. White, E.; Karp, C.; Strohecker, A.M.; Guo, Y.; Mathew, R. Role of autophagy in suppression of inflammation and cancer. *Curr. Opin. Cell Biol.* **2010**, *22*, 212–217. [CrossRef]
48. Galluzzi, L.; Bravo-San Pedro, J.M.; Kroemer, G. Autophagy Mediates Tumor Suppression via Cellular Senescence. *Trends Cell Biol.* **2016**, *26*, 1–3. [CrossRef]
49. Liu, H.; He, Z.; von Rutte, T.; Yousefi, S.; Hunger, R.E.; Simon, H.U. Down-regulation of autophagy-related protein 5 (ATG5) contributes to the pathogenesis of early-stage cutaneous melanoma. *Sci. Transl. Med.* **2013**, *5*, 202ra123. [CrossRef]
50. Young, A.R.; Narita, M.; Ferreira, M.; Kirschner, K.; Sadaie, M.; Darot, J.F.; Tavare, S.; Arakawa, S.; Shimizu, S.; Watt, F.M.; et al. Autophagy mediates the mitotic senescence transition. *Genes Dev.* **2009**, *23*, 798–803. [CrossRef]
51. Guo, J.Y.; Chen, H.Y.; Mathew, R.; Fan, J.; Strohecker, A.M.; Karsli-Uzunbas, G.; Kamphorst, J.J.; Chen, G.; Lemons, J.M.; Karantza, V.; et al. Activated Ras requires autophagy to maintain oxidative metabolism and tumorigenesis. *Genes Dev.* **2011**, *25*, 460–470. [CrossRef] [PubMed]
52. Wei, H.; Wei, S.; Gan, B.; Peng, X.; Zou, W.; Guan, J.L. Suppression of autophagy by FIP200 deletion inhibits mammary tumorigenesis. *Genes Dev.* **2011**, *25*, 1510–1527. [CrossRef] [PubMed]
53. Zhang, H.; Bosch-Marce, M.; Shimoda, L.A.; Tan, Y.S.; Baek, J.H.; Wesley, J.B.; Gonzalez, F.J.; Semenza, G.L. Mitochondrial autophagy is an HIF-1-dependent adaptive metabolic response to hypoxia. *J. Biol. Chem.* **2008**, *283*, 10892–10903. [CrossRef] [PubMed]
54. Amaravadi, R.; Kimmelman, A.C.; White, E. Recent insights into the function of autophagy in cancer. *Genes Dev.* **2016**, *30*, 1913–1930. [CrossRef] [PubMed]
55. Ferguson, L.R.; Chen, H.; Collins, A.R.; Connell, M.; Damia, G.; Dasgupta, S.; Malhotra, M.; Meeker, A.K.; Amedei, A.; Amin, A.; et al. Genomic instability in human cancer: Molecular insights and opportunities for therapeutic attack and prevention through diet and nutrition. *Semin Cancer Biol* **2015**, *35*, S5–S24. [CrossRef] [PubMed]
56. Kocaturk, N.M.; Akkoc, Y.; Kig, C.; Bayraktar, O.; Gozuacik, D.; Kutlu, O. Autophagy as a molecular target for cancer treatment. *Eur. J. Pharm. Sci.* **2019**, *134*, 116–137. [CrossRef]
57. Yang, Y.P.; Hu, L.F.; Zheng, H.F.; Mao, C.J.; Hu, W.D.; Xiong, K.P.; Wang, F.; Liu, C.F. Application and interpretation of current autophagy inhibitors and activators. *Acta Pharmacol Sin.* **2013**, *34*, 625–635. [CrossRef]
58. Mulcahy Levy, J.M.; Thorburn, A. Autophagy in cancer: Moving from understanding mechanism to improving therapy responses in patients. *Cell Death Differ.* **2019**. [CrossRef]
59. Ho, C.J.; Gorski, S.M. Molecular Mechanisms Underlying Autophagy-Mediated Treatment Resistance in Cancer. *Cancers* **2019**, *11*, 1775. [CrossRef]
60. Bursch, W.; Ellinger, A.; Kienzl, H.; Torok, L.; Pandey, S.; Sikorska, M.; Walker, R.; Hermann, R.S. Active cell death induced by the anti-estrogens tamoxifen and ICI 164 384 in human mammary carcinoma cells (MCF-7) in culture: The role of autophagy. *Carcinogenesis* **1996**, *17*, 1595–1607. [CrossRef]
61. Samaddar, J.S.; Gaddy, V.T.; Duplantier, J.; Thandavan, S.P.; Shah, M.; Smith, M.J.; Browning, D.; Rawson, J.; Smith, S.B.; Barrett, J.T.; et al. A role for macroautophagy in protection against 4-hydroxytamoxifen-induced cell death and the development of antiestrogen resistance. *Mol. Cancer Ther.* **2008**, *7*, 2977–2987. [CrossRef] [PubMed]
62. Schoenlein, P.V.; Periyasamy-Thandavan, S.; Samaddar, J.S.; Jackson, W.H.; Barrett, J.T. Autophagy facilitates the progression of ERalpha-positive breast cancer cells to antiestrogen resistance. *Autophagy* **2009**, *5*, 400–403. [CrossRef] [PubMed]
63. Lee, M.H.; Koh, D.; Na, H.; Ka, N.L.; Kim, S.; Kim, H.J.; Hong, S.; Shin, Y.K.; Seong, J.K.; Lee, M.O. MTA1 is a novel regulator of autophagy that induces tamoxifen resistance in breast cancer cells. *Autophagy* **2018**, *14*, 812–824. [CrossRef] [PubMed]
64. Paglin, S.; Hollister, T.; Delohery, T.; Hackett, N.; McMahill, M.; Sphicas, E.; Domingo, D.; Yahalom, J. A novel response of cancer cells to radiation involves autophagy and formation of acidic vesicles. *Cancer Res.* **2001**, *61*, 439–444. [PubMed]
65. Seglen, P.O.; Gordon, P.B. 3-Methyladenine: Specific inhibitor of autophagic/lysosomal protein degradation in isolated rat hepatocytes. *Proc. Natl. Acad. Sci. USA* **1982**, *79*, 1889–1892. [CrossRef] [PubMed]

66. Wu, Y.T.; Tan, H.L.; Shui, G.; Bauvy, C.; Huang, Q.; Wenk, M.R.; Ong, C.N.; Codogno, P.; Shen, H.M. Dual role of 3-methyladenine in modulation of autophagy via different temporal patterns of inhibition on class I and III phosphoinositide 3-kinase. *J. Biol. Chem.* **2010**, *285*, 10850–10861. [CrossRef]
67. Arico, S.; Petiot, A.; Bauvy, C.; Dubbelhuis, P.F.; Meijer, A.J.; Codogno, P.; Ogier-Denis, E. The tumor suppressor PTEN positively regulates macroautophagy by inhibiting the phosphatidylinositol 3-kinase/protein kinase B pathway. *J. Biol Chem* **2001**, *276*, 35243–35246. [CrossRef]
68. Maclean, K.H.; Dorsey, F.C.; Cleveland, J.L.; Kastan, M.B. Targeting lysosomal degradation induces p53-dependent cell death and prevents cancer in mouse models of lymphomagenesis. *J. Clin. Invest.* **2008**, *118*, 79–88. [CrossRef]
69. Rahim, R.; Strobl, J.S. Hydroxychloroquine, chloroquine, and all-trans retinoic acid regulate growth, survival, and histone acetylation in breast cancer cells. *Anticancer Drugs* **2009**, *20*, 736–745. [CrossRef]
70. Cerniglia, G.J.; Karar, J.; Tyagi, S.; Christofidou-Solomidou, M.; Rengan, R.; Koumenis, C.; Maity, A. Inhibition of autophagy as a strategy to augment radiosensitization by the dual phosphatidylinositol 3-kinase/mammalian target of rapamycin inhibitor NVP-BEZ235. *Mol. Pharmacol.* **2012**, *82*, 1230–1240. [CrossRef]
71. Sasaki, K.; Tsuno, N.H.; Sunami, E.; Tsurita, G.; Kawai, K.; Okaji, Y.; Nishikawa, T.; Shuno, Y.; Hongo, K.; Hiyoshi, M.; et al. Chloroquine potentiates the anti-cancer effect of 5-fluorouracil on colon cancer cells. *BMC Cancer* **2010**, *10*, 370. [CrossRef] [PubMed]
72. Sotelo, J.; Briceno, E.; Lopez-Gonzalez, M.A. Adding chloroquine to conventional treatment for glioblastoma multiforme: A randomized, double-blind, placebo-controlled trial. *Ann. Intern. Med.* **2006**, *144*, 337–343. [CrossRef] [PubMed]
73. Liu, Z.; Liu, J.; Li, L.; Nie, D.; Tao, Q.; Wu, J.; Fan, J.; Lin, C.; Zhao, S.; Ju, D. Inhibition of Autophagy Potentiated the Antitumor Effect of Nedaplatin in Cisplatin-Resistant Nasopharyngeal Carcinoma Cells. *PLoS ONE* **2015**, *10*, e0135236. [CrossRef] [PubMed]
74. Mauvezin, C.; Neufeld, T.P. Bafilomycin A1 disrupts autophagic flux by inhibiting both V-ATPase-dependent acidification and Ca-P60A/SERCA-dependent autophagosome-lysosome fusion. *Autophagy* **2015**, *11*, 1437–1438. [CrossRef] [PubMed]
75. Liu, J.; Xia, H.; Kim, M.; Xu, L.; Li, Y.; Zhang, L.; Cai, Y.; Norberg, H.V.; Zhang, T.; Furuya, T.; et al. Beclin1 controls the levels of p53 by regulating the deubiquitination activity of USP10 and USP13. *Cell* **2011**, *147*, 223–234. [CrossRef]
76. Shao, S.; Li, S.; Qin, Y.; Wang, X.; Yang, Y.; Bai, H.; Zhou, L.; Zhao, C.; Wang, C. Spautin-1, a novel autophagy inhibitor, enhances imatinib-induced apoptosis in chronic myeloid leukemia. *Int J. Oncol.* **2014**, *44*, 1661–1668. [CrossRef]
77. Mukhopadhyay, S.; Frias, M.A.; Chatterjee, A.; Yellen, P.; Foster, D.A. The Enigma of Rapamycin Dosage. *Mol. Cancer Ther.* **2016**, *15*, 347–353. [CrossRef]
78. Zhang, D.; Li, F.; Weidner, D.; Mnjoyan, Z.H.; Fujise, K. Physical and functional interaction between myeloid cell leukemia 1 protein (MCL1) and Fortilin. The potential role of MCL1 as a fortilin chaperone. *J. Biol. Chem.* **2002**, *277*, 37430–37438. [CrossRef]
79. Baylot, V.; Katsogiannou, M.; Andrieu, C.; Taieb, D.; Acunzo, J.; Giusiano, S.; Fazli, L.; Gleave, M.; Garrido, C.; Rocchi, P. Targeting TCTP as a new therapeutic strategy in castration-resistant prostate cancer. *Mol. Ther.* **2012**, *20*, 2244–2256. [CrossRef]
80. Bonhoure, A.; Vallentin, A.; Martin, M.; Senff-Ribeiro, A.; Amson, R.; Telerman, A.; Vidal, M. Acetylation of translationally controlled tumor protein promotes its degradation through chaperone-mediated autophagy. *Eur. J. Cell Biol.* **2017**, *96*, 83–98. [CrossRef]
81. Cuervo, A.M.; Wong, E. Chaperone-mediated autophagy: Roles in disease and aging. *Cell Res.* **2014**, *24*, 92–104. [CrossRef] [PubMed]
82. Kon, M.; Kiffin, R.; Koga, H.; Chapochnick, J.; Macian, F.; Varticovski, L.; Cuervo, A.M. Chaperone-mediated autophagy is required for tumor growth. *Sci. Transl Med.* **2011**, *3*, 109ra117. [CrossRef] [PubMed]
83. Gomes, L.R.; Menck, C.F.M.; Cuervo, A.M. Chaperone-mediated autophagy prevents cellular transformation by regulating MYC proteasomal degradation. *Autophagy* **2017**, *13*, 928–940. [CrossRef] [PubMed]
84. Kaushik, S.; Bandyopadhyay, U.; Sridhar, S.; Kiffin, R.; Martinez-Vicente, M.; Kon, M.; Orenstein, S.J.; Wong, E.; Cuervo, A.M. Chaperone-mediated autophagy at a glance. *J. Cell Sci.* **2011**, *124*, 495–499. [CrossRef]

85. Chiang, H.L.; Terlecky, S.R.; Plant, C.P.; Dice, J.F. A role for a 70-kilodalton heat shock protein in lysosomal degradation of intracellular proteins. *Science* **1989**, *246*, 382–385. [CrossRef]
86. Cuervo, A.M.; Dice, J.F. A receptor for the selective uptake and degradation of proteins by lysosomes. *Science* **1996**, *273*, 501–503. [CrossRef]
87. Kim, J.; Kundu, M.; Viollet, B.; Guan, K.L. AMPK and mTOR regulate autophagy through direct phosphorylation of Ulk1. *Nat. Cell Biol.* **2011**, *13*, 132–141. [CrossRef]
88. Bae, S.Y.; Byun, S.; Bae, S.H.; Min, D.S.; Woo, H.A.; Lee, K. TPT1 (tumor protein, translationally-controlled 1) negatively regulates autophagy through the BECN1 interactome and an MTORC1-mediated pathway. *Autophagy* **2017**, *13*, 820–833. [CrossRef]
89. Bommer, U.A.; Iadevaia, V.; Chen, J.; Knoch, B.; Engel, M.; Proud, C.G. Growth-factor dependent expression of the translationally controlled tumour protein TCTP is regulated through the PI3-K/Akt/mTORC1 signalling pathway. *Cell Signal* **2015**, *27*, 1557–1568. [CrossRef]
90. Kim, Y.C.; Guan, K.L. mTOR: A pharmacologic target for autophagy regulation. *J. Clin. Invest.* **2015**, *125*, 25–32. [CrossRef]
91. Liang, X.H.; Kleeman, L.K.; Jiang, H.H.; Gordon, G.; Goldman, J.E.; Berry, G.; Herman, B.; Levine, B. Protection against fatal Sindbis virus encephalitis by beclin, a novel Bcl-2-interacting protein. *J. Virol.* **1998**, *72*, 8586–8596. [CrossRef] [PubMed]
92. Kang, R.; Zeh, H.J.; Lotze, M.T.; Tang, D. The Beclin 1 network regulates autophagy and apoptosis. *Cell Death Differ.* **2011**, *18*, 571–580. [CrossRef] [PubMed]
93. Pattingre, S.; Tassa, A.; Qu, X.; Garuti, R.; Liang, X.H.; Mizushima, N.; Packer, M.; Schneider, M.D.; Levine, B. Bcl-2 antiapoptotic proteins inhibit Beclin 1-dependent autophagy. *Cell* **2005**, *122*, 927–939. [CrossRef] [PubMed]

cells MDPI

Article

Mmi1, the Yeast Ortholog of Mammalian Translationally Controlled Tumor Protein (TCTP), Negatively Affects Rapamycin-Induced Autophagy in Post-Diauxic Growth Phase

Jana Vojtova * and Jiri Hasek *

Institute of Microbiology of the Czech Academy of Sciences, Videnska 1083, Prague 4 142 20, Czech Republic
* Correspondence: jana.vojtova@biomed.cas.cz (J.V.); hasek@biomed.cas.cz (J.H.); Tel.: +420-241-062-504 (J.V.); +420-241-062-343 (J.H.)

Received: 15 November 2019; Accepted: 3 January 2020; Published: 7 January 2020

Abstract: Translationally controlled tumor protein (TCTP) is a multifunctional and highly conserved protein from yeast to humans. Recently, its role in non-selective autophagy has been reported with controversial results in mammalian and human cells. Herein we examine the effect of Mmi1, the yeast ortholog of TCTP, on non-selective autophagy in budding yeast *Saccharomyces cerevisiae*, a well-established model system to monitor autophagy. We induced autophagy by nitrogen starvation or rapamycin addition and measured autophagy by using the Pho8Δ60 and GFP-Atg8 processing assays in WT, *mmi1Δ*, and in autophagy-deficient strains *atg8Δ* or *atg1Δ*. Our results demonstrate that Mmi1 does not affect basal or nitrogen starvation-induced autophagy. However, an increased rapamycin-induced autophagy is detected in *mmi1Δ* strain when the cells enter the post-diauxic growth phase, and this phenotype can be rescued by inserted wild-type *MMI1* gene. Further, the *mmi1Δ* cells exhibit significantly lower amounts of reactive oxygen species (ROS) in the post-diauxic growth phase compared to WT cells. In summary, our study suggests that Mmi1 negatively affects rapamycin-induced autophagy in the post-diauxic growth phase and supports the role of Mmi1/TCTP as a negative autophagy regulator in eukaryotic cells.

Keywords: Mmi1; TCTP; translationally controlled tumor protein; autophagy; reactive oxygen species; rapamycin; nitrogen starvation

1. Introduction

TCTP (Translationally Controlled Tumor Protein) is an evolutionarily-conserved and abundant protein among eukaryotic organisms. It is an essential protein for the development of multicellular organisms [1–3] and its main biological role is likely an anti-apoptotic activity [4–7]. However, it is also involved in many other core cell biological processes (reviewed in [8]). Despite it being a long time since its discovery in the 1980s [9] and subsequent intensive studies, the protein still remains a bit enigmatic, and new discoveries and new effects are still being described.

Recently, TCTP has also been found to affect autophagy [10,11]. Chen and colleagues reported that TCTP positively affects hypoxia and starvation-induced bulk non-selective autophagy [10]. On the other hand, Bae and colleagues have declared TCTP a negative regulator of basal and rapamycin-induced non-selective autophagy [11]. Since TCTP is an evolutionarily highly conserved protein, we used a pioneer model organism for studying autophagy, budding yeast *Saccharomyces cerevisiae* [12], reviewed in [13–15] to test the effect of yeast TCTP on autophagy. The *S. cerevisiae* is usually batch cultured and its growth in the culture is highly affected by a carbon source. When glucose is added to yeast cells, they rapidly adapt to fermentation of the rich carbon source during a short lag-phase. After the adaptation, they start to ferment the sugar and reach a maximal growth rate. This phase is called the exponential growth phase. Once glucose becomes limiting, yeast cells enter a second lag-phase, known as a diauxic shift [16]. During the diauxic shift yeast cells change their metabolism from fermentation to respiration. The diauxic shift is followed by a slow growing phase (post diauxic growth phase), during which ethanol, acetate, and other fermentation products are utilized by respiration. When the all carbon sources are exhausted yeast cells enter a quiescence or stationary phase (G_0) [16,17].

Yeast TCTP was originally described as a translation machinery associated protein, Tma19 [18]. Later, Tma19 was described as a microtubule and mitochondria interacting protein and renamed Mmi1 [19]. Therefore, we refer to the protein as Mmi1 hereafter. Mmi1 is a small 18.7 kDa, acidic (pI = 4.17), and highly abundant protein in exponentially growing yeast cells corresponding to approximately 200,000 molecules per cell [20]. During the post-diauxic growth phase, Mmi1 is still a highly abundant protein exhibiting a steady-state level of expression [21], and its abundance continually decreases in the stationary phase [21,22]. Upon rapamycin treatment, the Mmi1 protein pool decreases [23], indicating that the Mmi1 expression in yeast might be regulated by TOR pathway similarly to higher eukaryotic cells [24]. Further, the *mmi1*Δ strain exhibits a slow growth phenotype [19], indicating that Mmi1 is a pro-survival factor. Mmi1 is uniformly distributed in cytosol, but if stress is applied, its distribution is changed. Upon mild oxidative stress, Mmi1 translocates to mitochondria [25], while upon heat stress it relocalizes to the nucleus and mitochondria and is also present in stress granules [26]. Mmi1 role in the nucleus is not clarified yet. However, recently Bischof and colleagues suggested a model that the mitochondrial localization of the Mmi1/TCTP is responsible for the clearance of the mitochondrial membrane from harmful proteins in a time of stress [25], thereby protecting cells from apoptosis. Above the anti-apoptotic function, Mmi1 affects a wide range of biological functions and processes most likely through interaction with its binding partners. According to the BioGRID database [27], Mmi1 currently possess about 49 physically interacting protein partners. These proteins are mainly involved in cell cycle, transcription, translation, and protein degradation. Indeed, our previous results indicated that Mmi1 modulates activity of proteasomes [26], the major protein degradation system in all eukaryotic cells next to autophagy. Nevertheless, the effect of Mmi1 on autophagy in yeast cells has not been tested yet. To test the question of whether Mmi1 affects non-selective autophagy, we induced autophagy through different conditions and used independent approaches to monitor the autophagy.

Autophagy (here referred to macroautophagy) occurs constitutively at basal levels, but it is dramatically stimulated by starvation and by various stresses [28,29]. It allows cells to respond to various types of stresses and to adapt to changing nutrient conditions [30]. Autophagy can be either a non-selective self-consumption or a selective consumption of specific cargoes or organelles. The bulk autophagy is completely inhibited in nutrient-rich conditions, but can be induced by shifting cells to starvation medium [31] or by addition of rapamycin [32], a potent inhibitor of TORC1 (target of rapamycin complex 1) [33,34]. During non-selective autophagy a portion of cytosol is sequestered for degradation into double-membrane structures named autophagosomes, which are delivered to the vacuole and degradated by vacuolar hydrolases [35]. In *S. cerevisiae*, eighteen Atg proteins, Atg1–10, Atg12–14, Atg16–18, Atg29 and Atg31 play essential roles in autophagy, and these core proteins are required for the formation of autophagosomes (reviewed in [15,36]). When non-selective autophagy is induced, the Atg17-Atg29-Atg31 complex act as an essential scaffold that facilitates formation of the preautosomal structure (PAS), from which the autophagosome is generated [37]. Autophagy is involved in a variety of physiological processes. In unicellular eukaryotes it takes care of cellular housekeeping and sustaining viability, and it is also essential for adaptation to a new host and formation of spores [38]. In higher eukaryotes it is important for cell survival and maintenance, and its dysfunction contributes to the pathologies of many diseases, e.g., cancer [39].

Here, we examined the effect of Mmi1 on bulk non-selective autophagy in yeast. Our results demonstrate a negative effect of Mmi1 on rapamycin-induced autophagy in contrast to nitrogen starvation-induced autophagy. Interestingly, the negative effect of Mmi1 on rapamycin-induced autophagy is detected after diauxic shift.

2. Materials and Methods

2.1. Yeast Strains, Media, and Growth Conditions

Yeast strains used in this study are listed in Table 1. Yeast cells were grown in shaking flask at 30 °C in rich medium (YPD; 1% *w*/*v* yeast extract, 2% *w*/*v* peptone, 2% *w*/*v* glucose) or defined synthetic medium (SD; 0.17% *w*/*v* yeast nitrogen base, 0.5% *w*/*v* ammonium sulfate, 2% *w*/*v* glucose, and auxotrophic amino acids as required). Starvation experiments were performed in synthetic minimal medium lacking nitrogen (SD-N; 0.17% *w*/*v* yeast nitrogen base without ammonium sulfate and amino acids, 2% *w*/*v* glucose).

Table 1. Yeast strains used in this study.

Strain	Relevant Genotype	Source
CRY155	BY4741; *MATa, his3Δ1 leu2Δ0 met15Δ0 ura3Δ0*	[40]
CRY1107	BY4741; *MATa mmi1::KanMX4*	Euroscarf
CRY2829	*MATa, leu2–3,112, trp1, ura3–52, pho8::pho8Δ60, pho13::URA3*	[41]
CRY2830	*MATa, leu2–3,112, trp1, ura3–52, pho8::pho8Δ60, pho13::URA3, atg8::KanMX*	[42]
CRY2833	*MATa, leu2–3,112, trp1, ura3–52, pho8::pho8Δ60, pho13::URA3, mmi1::natNT2*	This study
CRY2645	BY4741; *MATa [pRS316GFPAut7]*	This study
CRY2673	BY4741; *MATa, mmi1::KanMX4 [pRS316GFPAut7]*	This study
CRY2662	BY4741; *MATa, atg1::natNT2*	This study
CRY2665	BY4741; *MATa, atg1::natNT2 [pRS316GFPAut7]*	This study
CRY2959	MATa, *leu2–3,112, trp1, ura3–52, pho8::pho8Δ60, pho13::URA3, mmi1::natNT2 pAG32-MMI1 (hphMX6)*	This study

2.2. Plasmids

The pAG32-*MMI1* plasmid was constructed by inserting a fragment containing 348 bps upstream of the *MMI1* ORF, *MMI1* ORF, and 454 bps downstream of *MMI1* ORF into the SacI and SpeI sites of the pAG32 plasmid. The fragment was amplified by PCR from pRS316-*MMI1* plasmid. Before the transformation of yeast cells, the plasmid was cut with BsaAI restriction enzyme at the position of 319 bps downstream of *MMI1* ORF.

The pRS316-*MMI1* plasmid was made through PCR cloning. The DNA fragment containing 500 base pairs (bps) upstream of the start codon together with the *MMI1* ORF and 500 bps downstream of the ORF were amplified by PCR from WT yeast genomic DNA. The fragment was then ligated into the SacI and EcoRI sites of pRS316 vector, resulting in the pRS316-*MMI1* plasmid.

2.3. Phosphatase Assay

Pho8Δ60-expressing strains (WT, *mmi1Δ* and *atg8Δ*) were grown in YPD medium to early log phase ($OD_{600} \approx 0.8$) and then shifted to SD-N medium (after double wash with H_2O and one wash with SD-N) or rapamycin was added to final concentration 200 nM. The cells were collected at indicated time points, and the alkaline phosphatase activity of Pho8Δ60 was carried out as described previously [41,43,44]. In total, five OD_{600} equivalents of yeast cells were harvested, washed once with ice-cold water, and once with wash buffer (0.85% *w*/*v* NaCl and 1mM PMSF) and resuspended in 500 μL lysis buffer (20 mM Pipes, pH 7.0, 0.5% *v*/*v* Triton X-100, 50 mM KCl, 100 mM potassium acetate, 10 mM MgSO4, 10 μM ZnSO4, and 1 mM PMSF). The cells were lysed with 250 μL equivalents of glass beads using a Fast-prep desintegrator 5 times for 20 s at 4 °C and incubated for 2 min on ice in-between. The lysates were centrifuged at 14,000× *g* for 5 min at 4 °C. The supernatant was collected and 100 μL of the supernatant was added to 400 μL reaction buffer (250 mM Tris-HCl, pH 8.5, 0.4% *v*/*v* Triton X-100, 10 mM MgSO4, and 1.25 mM p-nitrophenyl phosphate). Samples were incubated for 10 min at 30 °C before terminating the reaction by adding 500 μL of stop buffer (2 M glycine, pH 11). Production of p–nitrophenol was monitored by measuring the absorbance at 400 nm (A_{400}) using a spectrophotometer (Helios Gamma Spectrophotometer, Unicam), and the concentration in nmol of p–nitrophenol in the samples was calculated from a standard curve of commercial p–nitrophenol (Sigma) (0 to 100 nmol). Protein concentration in the extracts was measured with the Pierce™ BCA Protein Assay Kit (Thermo Scientific, Rockford, IL, USA), and the specific activity was calculated as nmol p–nitrophenol/min/mg protein. The statistical evaluation of phosphatase activity was performed by two-way analysis of variance (ANOVA) using R software.

2.4. *Quantification of Glucose*

Yeast cultures were centrifuged, supernatants were collected, and glucose concentration was measured by GLU 500 kit (Erba Lachema s.r.o., Brno, Czech Republic) according to the manufactured conditions.

2.5. *GFP-Atg8 Processing Assay*

BY4741 WT, *mmi1*Δ, and *atg1*Δ strains transformed with plasmid pRS316_GFPAUT7 [45] carrying GFP-Atg8 were grown in SC medium to $OD_{600} \approx 0.8$ and then shifted to SD-N medium (after double wash with H_2O and one wash with SD-N) or rapamycin was added to final concentration 200 nM. In indicated time points, samples were collected and normalized to $OD_{600} \approx 1$ and 100% TCA was added to the final concentration of 12.5%. Samples were frozen at −80 °C for at least 1 h, centrifuged for 5 min at 15,000× *g*, and pellets were washed with an ice-cold 80% aceton and dried at room temperature. Dried pellets were resuspended in 50 µL of 0.1N NaOH and 1% *w*/*v* SDS and bath sonicated. Then 50 µL of 2 × SDS loading buffer (100 mM Tris-HCl pH 6.8, 4% *w*/*v* SDS, 20% *v*/*v* glycerol, 0.2% *w*/*v* bromphenol blue) with 0.1M DDT was added and the lysates were resolved by 10% SDS-PAGE and transferred to Protran nitrocellulose membrane (Sigma-Aldrich/Merck, St. Louis, MO, USA). The membrane was blocked with 5% *w*/*v* non-fat dried milk (Regilait, Macon, France) and incubated overnight with a mouse monoclonal anti-GFP antibody conjugated with horseradish peroxidase (sc-9996, Santa Cruz, USA) at 1:1000. As a loading control, detection of Pgk1 with anti-Pgk1 antibody (Abcam, ab113687, 1:10000) and goat anti-mouse IgG conjugated with horseradish peroxidase (Thermo Fisher Scientific, No: 32430, 1:1000) was used. Proteins were detected by SuperSignal™ West Dura Substrate (Thermo Scientific, Rockford, IL, USA). In order to calculate the ratio of free GFP to Pgk1, Western blot signals were detected by G:BOX Chemi-XRQ gel documentary system (Syngene, Cambridge, UK) and quantified by ImageJ program [46].

2.6. *Fluorescence Microscopy*

Fluorescence microscopy was performed using an Olympus IX-81 inverted microscope equipped with a motorized stage, a 100× PlanApochromat oil-immersion objective (NA = 1.4) and a Hamamatsu Orca-ER-1394 digital camera. Images were processed using Olympus Cell-R™ Xcellence, Adobe CS5 software packages. The same exposure time was used to detect GFP signal in all tested strains. To quantify the cellular distribution of GFP-Atg8 signal, cells were co-stained with a vacuolar marker FM4-64 (1 µg/mL, 1 h, 30 °C) and scored into three categories: vacuole & cytosol, cytosol, and vacuole. In the category "vacuole and cytosol" the GFP signal was evenly distributed within the cells and there were no signal gradients between vacuole and cytosol. In the category "cytosol" the higher GFP signal was detected in cytosol compared to vacuole. Finally, in the category "vacuole" the higher GFP signal was detected in vacuole compared to cytosol. The cells with oversaturated signal were excluded from the quantification.

2.7. *Oxygen Radicals' Measurement by Dihydroethidium Staining*

Dihydroethidium (DHE) staining was performed as described in Neklesa and Davis, 2008 [47]. In total 1×10^7 cells/mL were stained with 15 µg/mL of dihydroethidium (DHE) in YPD for 1 h at 30 °C under shacking. Cells were washed in PBS, reinoculated in PBS, and in total 20,000 cells were analyzed in a Flow Cytometer BD LSRII (BD Biosciences, USA). To detect specifically the DHE oxidation product hydroxyethidium, an excitation wavelength of 405 nm was used. This wavelength is the closest available wavelength to a distinct hydroxyethidium excitation maximum of 396 nm that is not present for other DHE oxidation products [48]. The signal was collected by using emission filter 576/26, and cell viability was evaluated by Hoechst 33,258 (1 μg/mL). The data were analyzed by FlowJo software and mean fluorescence DHE intensity values were used to compare the total DHE fluorescence of the strains.

2.8. Viability Assay for Testing Rapamycin Sensitivity

To compare the sensitivity of WT and *mmi1*Δ strains to rapamycin, we incubated exponentially growing cells in 24 well plate in YPD with 0, 5, 6, 8, 10, and 15 nM of rapamycin. The cells were seeded to initial OD ≈ 0.1 and incubated overnight at 30 °C in Eon^{TM} Miroplate Spectrophotometer (BioTek) under agitation. Every 15 min OD_{600} was determined in each well by Gen5 Microplate Reader and Imager Software. Due to increased concentration of rapamycin were obtained outgrowth curves with a distinct shift in the curves as cells lose viability. For each concentration time shift for OD ≈ 0.5 in the outgrowth curve relative to the curve of untreated cells was determined and viability curves were calculated as described in details in [49].

3. Results

3.1. Basal and Nitrogen Starvation-Induced Autophagy are not Affected by Mmi1

A previous study in mammalian cells indicated that TCTP promotes autophagy under hypoxia and starvation conditions [10]. Later, however, the results were challenged by a study demonstrating TCTP as a negative regulator of non-selective autophagy [11]. To investigate the role of the yeast TCTP ortholog, Mmi1, in non-selective autophagy, we examined autophagy in wild-type (WT) and *mmi1*Δ cells upon shift to a nitrogen starvation medium that stimulates autophagy induction [12]. As a negative control we used autophagy deficient strains *atg8*Δ depleted for the key autophagy molecule Atg8 [50]. The non-selective autophagy was determined by the quantitative Pho8Δ60 assay [43]. The Pho8Δ60 assay is an enzymatic assay that utilizes a truncated version of the alkaline phosphatase Pho8Δ60 that lacks the targeting sequence to endoplasmic reticulum, and thus remains in the cytosol. Upon autophagy induction, Pho8Δ60 is delivered to the vacuole, gets activated by the proteolytic cleavage, and serves as a marker of the amount of cytosol delivered through the non-selective autophagy [43]. Yeast strains were grown in the YPD medium until the early logarithmic phase, washed, and shifted to the nitrogen starvation (SD-N) medium. As shown in Figure 1A, at 0 h upon shift to SD-N media low values of the phosphatase activity were detected in all tested strains, indicating that basal levels of autophagy were not affected. Upon prolonged nitrogen starvation, phosphatase activities increased. However, similar values were detected for the WT and the *mmi1*Δ strains at all tested time points. On the other hand, the control *atg8*Δ strain exhibited only a low level of autophagy, demonstrating that delivery of Pho8Δ60 to the vacuole depends on autophagy. These results indicated that Mmi1 did not affect the nitrogen starvation-induced non-selective autophagy.

To confirm the results by an independent approach we also performed GFP-Atg8 processing assay [28]. The GFP-Atg8 assay is based on Western blot detection of a free GFP moiety released from a core autophagy protein Atg8 that is resistant to vacuolar proteases [28]. As a negative control we used autophagy deficient *atg1*Δ depleted for the key autophagy molecule Atg1 [51]. As shown in Figure 1B, no free GFP band was detected by Western blot at 0 h time point of nitrogen starvation in WT, *mmi1*Δ, and *atg1*Δ strain demonstrating that basal level of autophagy was not affected in the strains. Upon prolonged nitrogen starvation, similar levels of the free GFP band were detected in the WT and the *mmi1*Δ cell lysates, indicating that Mmi1 did not influence GFP-Atg8 cleavage upon nitrogen starvation. Further, no free GFP band was detected in the negative control *atg1*Δ strain at all tested time points, demonstrating that GFP-Atg8 cleavage was dependent on the autophagic degradation. These results, together with the results obtained by the phosphatase assay, demonstrate that the basal and the nitrogen starvation-induced autophagy are not influenced by Mmi1 in yeast *S. cerevisiae*.

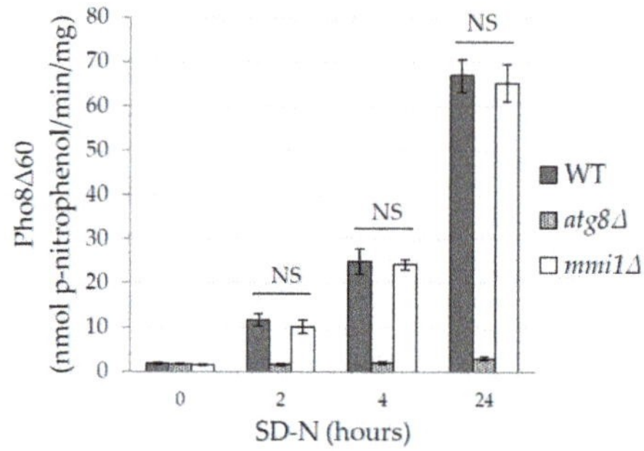

(A)

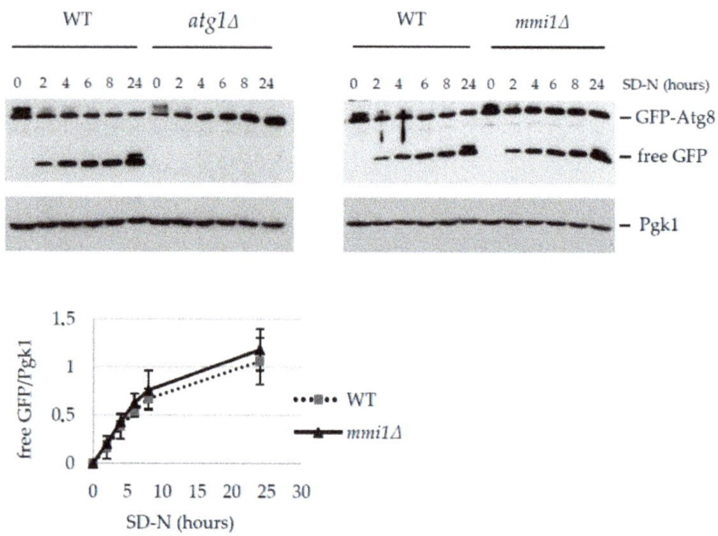

(B)

Figure 1. Basal and nitrogen starvation-induced autophagy normally occur in *mmi1*Δcells. (**A**) Exponentially growing WT, *atg8*Δ, and *mmi1*Δcells expressing Pho8Δ60 ($OD_{600} \approx 0.8$) were shifted to nitrogen starvation medium (SD-N). Samples were taken in indicated time points, proteins extracted, and the specific Pho8Δ60 activity was determined. Results are means ± SD of three independent experiments performed in duplicates (n = 6). The statistical evaluation was performed by using two way analysis of variance (ANOVA). The threshold for significance was set as $p \leq 0.01$. NS; not significant; (**B**) Western blot detection of GFP-Atg8 cleavage in the WT, *atg1*Δ, and *mmi1*Δcells. Cells expressing GFP-Atg8 were grown until the logarithmic growth phase ($OD_{600} \approx 0.8$) and then shifted to the SD-N medium.. Samples were taken at indicated time points and the cleavage of GFP-Atg8 was analyzed by Western blot detection with antibodies against GFP. Detection of Pgk1 was used as a loading control. Quantification of the blots is presented below. GFP ratio (free GFP/Pgk1) was calculated. Error bars reflect SD from the two independent experiments.

3.2. Mmi1 Negatively Affects Rapamycin-Induced Autophagy When the Cells Enter Post-Diauxic Growth Phase

A previous study on HeLa cells demonstrated a negative effect of TCTP on rapamycin-induced non-selective autophagy [11]. Therefore, we also examined the effect of Mmi1 on autophagy induced by rapamycin, a potent inhibitor of the TOR pathway [33,34] that stimulates autophagy [32]. Yeast cells were grown in rich YPD medium until the early logarithmic growth phase. Autophagy was induced by the addition of rapamycin and determined by the phosphatase assay. As shown in Figure 2A, very low autophagy levels were detected at 0 h after rapamycin addition in all strains, demonstrating consistently with our previous results that the basal autophagy is not affected by Mmi1. After prolonged incubation with rapamycin, autophagy levels increased but no difference between the WT and the *mmi1*Δ strains was detected 2 and 4 h after the autophagy induction. However, an increased autophagy was detected 18 and 24 h after rapamycin addition to *mmi1*Δ cells (Figure 2A). Low levels of autophagy were detected in the control *atg8*Δ cells in all tested time points. These results indicated that rapamycin-induced autophagy was increased in the *mmi1*Δ strain after a long term incubation with rapamycin.

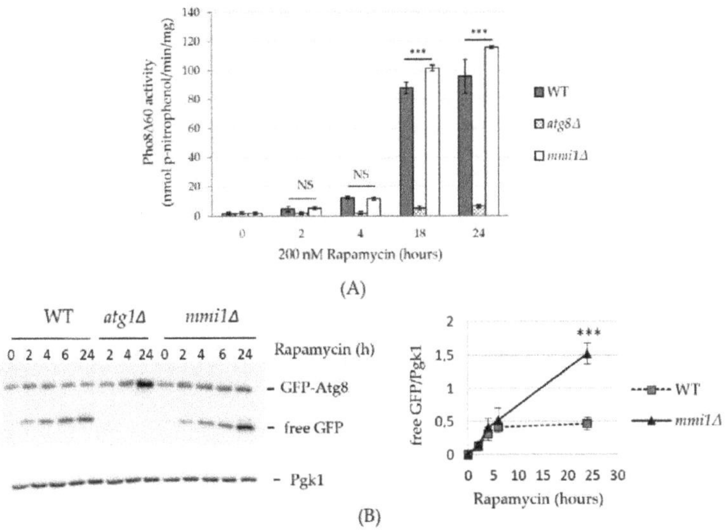

Figure 2. Non-selective autophagy is promoted in the *mmi1*Δ strain after prolonged incubation with rapamycin. (**A**) Exponentially growing WT, *atg8*Δ, and *mmi1*Δ cells expressing Pho8Δ60 ($OD_{600} \approx 0.8$) were treated with rapamycin 200 nM. Samples were taken at indicated time points, proteins extracted, and the specific Pho8Δ60 activity was determined. Results are means ± SD of two independent experiments performed in duplicates (n = 4). The statistical evaluation was performed by using two way analysis of variance (ANOVA). NS; not significant; *** $p < 0.001$. (**B**) Western blot detection of GFP-Atg8 cleavage in WT, *atg1*Δ, and *mmi1*Δ cells after rapamycin addition. Cells expressing GFP-Atg8 were grown until log phase (OD ≈ 0.8) and then rapamycin was added to a final concentration 200 nM. At indicated time points proteins were extracted and the protein lysates were examined by Western blot. Pgk1 was used as a loading control. Quantification of the blots is presented to the right. GFP ratio (free GFP/Pgk1) was calculated. *** $p < 0.01$. Error bars reflects SD from three independent experiments.

To corroborate the results, we also performed GFP-Atg8 processing assays. Consistently, similar levels of the free GFP were detected in the WT and *mmi1*Δ cell lysates at early time points upon autophagy induction, and no free GFP was detected in the control *atg1*Δ cell lysates at all tested time points (Figure 2B). However, an increased accumulation of the free GFP was detected in the *mmi1*Δ cell lysate 24 hours after rapamycin addition. The increased GFP-Atg8 processing in *mmi1*Δ strain 24 hours upon rapamycin addition could be also confirmed by fluorescence microscopy (Figure 3). The majority of the GFP-Atg8 signal in the *mmi1*Δ cells was present in the vacuole compared to WT cells, while in the *atg1*Δ cells the majority of the GFP signal was present outside of the vacuole (Figure 3A, B). These results, together with the results of the phosphatase assay demonstrate the increased autophagy in the *mmi1*Δ strain at later time points after rapamycin addition.

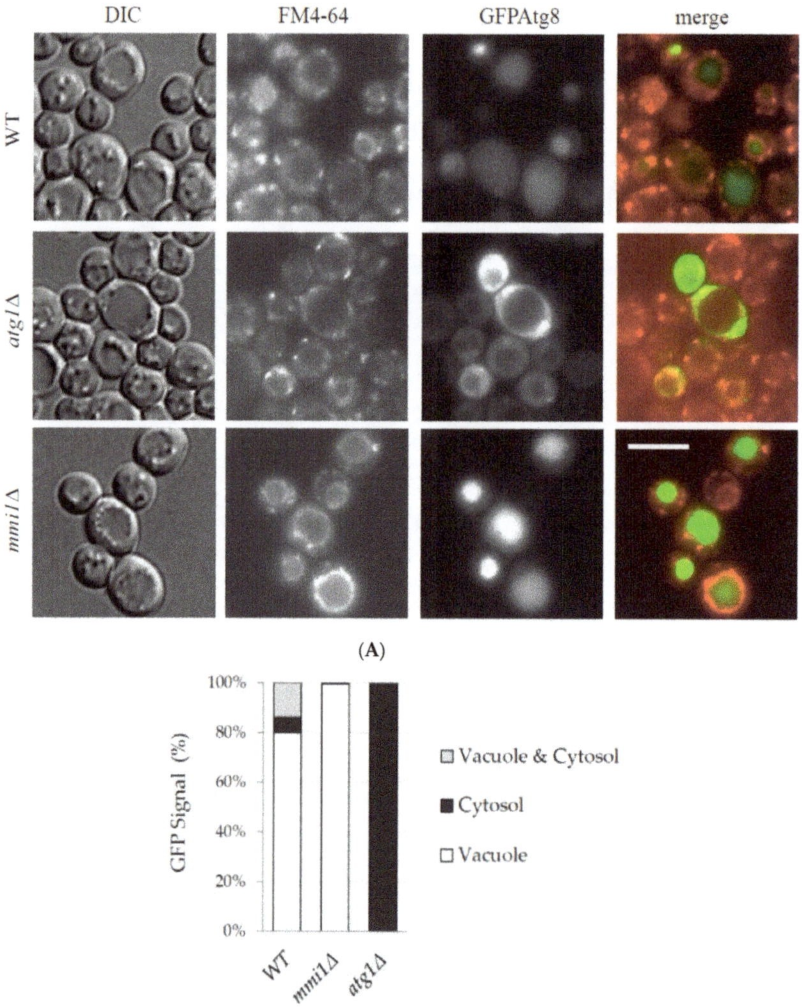

Figure 3. Fluorescence microscopy detection of a higher GFP-Atg8 processing in *mmi1*Δ cells upon rapamycin treatment. (**A**) The same samples of WT, *atg1*Δ, and *mmi1*Δ cells treated with rapamycin for 24 h and analyzed by Western blot in Figure 2B were examined under fluorescence microscope. Vacuoles were labelled by the red fluorescence probe FM4-64 (1 μg/mL, 1h) and GFP signal was taken under the same exposure time in all tested strains. DIC, differential interference contrast. Scale bar, 5 μm. (**B**) Quantification of the GFP-Atg8 cellular distribution 24 h after rapamycin addition in WT, *mmi1*Δ, and *atg1*Δ strains. Results are means from two independent experiments and each bar represents 250 cells.

To verify that the phenotype seen in the *mmi1*Δ mutant strain was not due to an unknown secondary mutation, we created a new strain (CRY2959) possessing the wild-type *MMI1* gene under control of its endogenous promotor in the *mmi1*Δ strain. As shown in Figure 4, the presence of *MMI1* wild-type gene decreased the autophagy activity of *mmi1*Δ strain to the WT strain, suggesting that the mutant phenotype is indeed the result of deleting the *MMI1* gene.

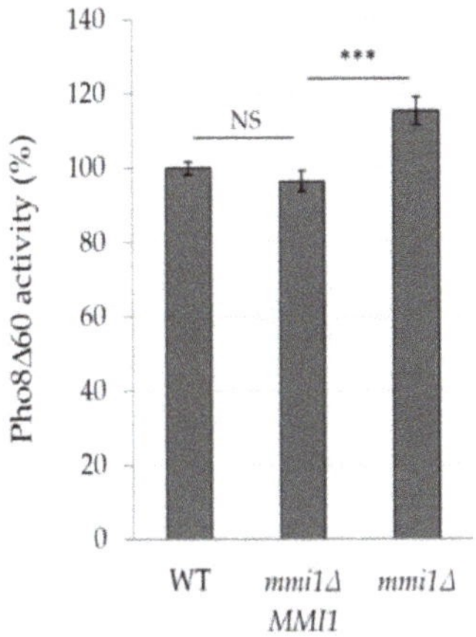

Figure 4. The increased rapamycin-induced autophagy in the *mmi1*Δ strain can be rescued by wild-type *MMI1* gene. Exponentially growing cells of the WT, *mmi1*Δ, and *mmi1*Δ *MMI1* strains expressing Pho8Δ60 ($OD_{600} \approx 0.8$) were treated with 200 nM rapamycin for 24 h and the specific Pho8Δ60 was determined. Results are expressed as means ± SD from two independent experiments performed in duplicates (n = 4). The statistical evaluation was performed by using two way analysis of variance (ANOVA). *** $p < 0.001$; NS; not significant.

Our results demonstrated the increased autophagy in *mmi1*Δ cells after a long-term incubation with rapamycin. However, no similar effect was detected in the case of the nitrogen starvation-induced autophagy. Nitrogen starvation and rapamycin treatment are widely used approaches to induce autophagy in yeast [31,32]. In the nitrogen starvation conditions, yeast cells complete division and arrest in the G1/G2 quiescence state [52], increase their volume due to enlarged vacuole by autophagy induction, and remain viable for two days [12]. Rapamycin treatment is believed to mimic nutrient deprivation, including greatly reduced cell growth [33] and autophagy induction [32]. However, it has been demonstrated independently by several groups that even at high concentration of rapamycin, yeast cells maintain their proliferative ability [47,53–56]. We used batch cultivation, and cells were grown in rich YPD media to $OD_{600} \approx 0.8$ before rapamycin was added to a final concentration 200 nM, or cells were washed and shifted to nitrogen starvation media. Both approaches induced autophagy as shown earlier (see Figures 1 and 2). Nevertheless, as shown in Figure 5, rapamycin and nitrogen starvation-treated cells highly differed in the cell growth. In nitrogen starvation medium the WT and the *mmi1*Δ strains reached only $OD_{600} \approx 2$. In fact, the OD_{600} value in the SD-N medium seems to be overestimated likely due to an increased size of the cells. Indeed, when the number of cells per ml was measured, the increase of about 20% only was detected (Supplementary Figure S1). On the other hand, WT and *mmi1*Δ strain treated with 200 nM rapamycin exhibited higher growth rates and the cell cultures reached $OD_{600} \approx 7$ after 24 h of cultivation after rapamycin addition. These results demonstrate a striking difference between used nitrogen starvation conditions and rapamycin treatment in batch-cultivated yeast cells. To characterize the cell growth during nitrogen starvation and rapamycin treatment, we measured concentration of glucose in media. As showed in Figure 5B, glucose was still present even 24 h upon the shift to the SD-N media. However, glucose was completely exhausted in the YPD media at latest 15 h after rapamycin addition (Figure 5B). Since *S. cerevisiae* cultivated in the glucose-containing media undergoes a growth arrest (diauxic shift) after glucose depletion, no presence of glucose in the media demonstrates that the cells have already entered the post diauxic growth phase. To correlate glucose depletion with autophagy induction, we measured phosphatase activity in indicated time points I and II as shown in Figure 5A,B. The time point I represented an exponential phase, where glucose was present in media and the time point II represented the post-diauxic growth phase; where also glucose was absent from the media (Figure 5B). Our results indicate that autophagy is promoted in the *mmi1*Δ strain after rapamycin treatment, when glucose is already not present in the media (Figure 5C). These results altogether demonstrate that the rapamycin-induced autophagy is increased in the *mmi1*Δ strain after diauxic shift.

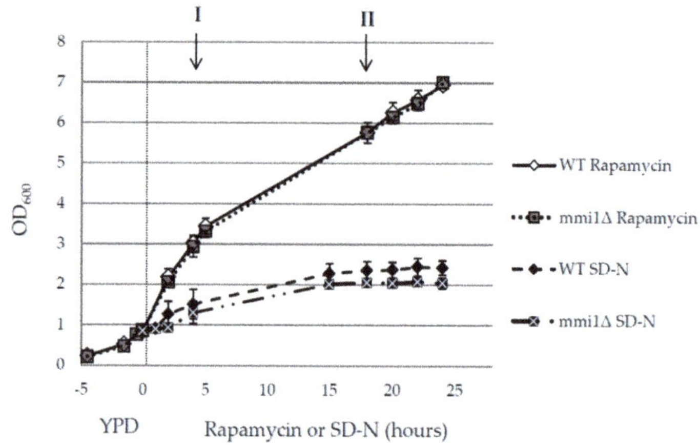

(A)

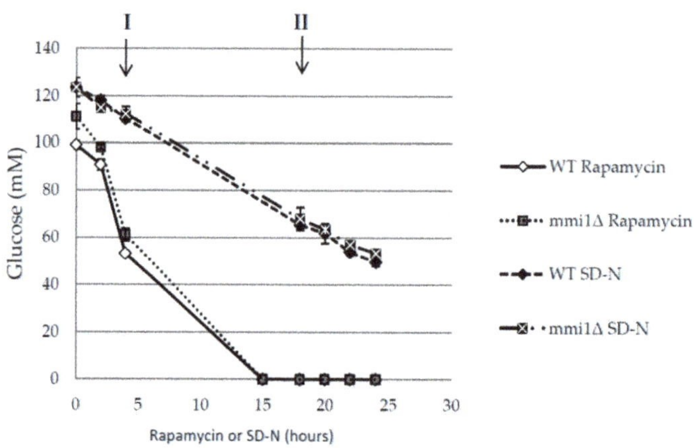

(B)

Figure 5. *Cont.*

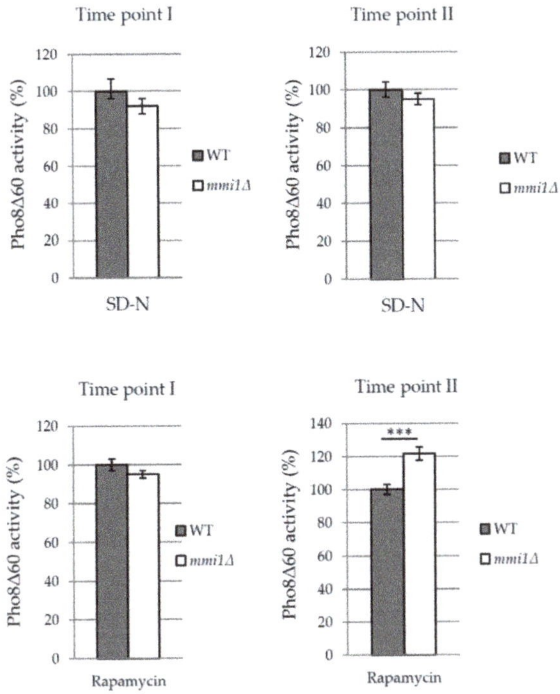

(C)

Figure 5. Increased rapamycin-induced autophagy in *mmi1*Δ strain occurs after glucose exhaustion. (**A**) Exponentially growing WT and *mmi1*Δ cells expressing Pho8Δ60 ($OD_{600} \approx 0.8$) in the YPD medium were either shifted to the SD-N medium or treated with rapamycin (200 nM). Optical density at 600 nm was measured at indicated time points. Results are presented as means ± SD of three independent experiments performed in duplicates (n = 6). (**B**) Concentration of glucose in the media as shown in A was measured after the addition of rapamycin (200 nM) to WT and *mmi1*Δ strains or after the shift of the strains to SD-N media. Results are means of two independent experiments performed in triplicates (n = 6). (**C**) Pho8Δ60 assay was measured in time points I and II as indicated in A, and B. Point I represents the exponential growth phase where glucose is still present in the medium. Point II represents the post-diauxic phase where glucose is already exhausted. Results are normalized to the WT strain (100%) and represent means ± SD from two independent experiments performed in duplicates (n = 4). The statistical evaluation was performed by using two way analysis of variance (ANOVA). *** $p < 0.001$.

3.3. *In Post Diauxic Growth Phase Amount of Superoxide Radicals is Decreased in mmi1Δ Strain*

Our results demonstrated that Mmi1 negatively affects the rapamycin-induced autophagy in the post diauxic growth phase. Rapamycin forms a complex with FKBP (Fpr1 in yeast) to inhibit TORC1 [33]. Neklesa and Davis reported that superoxide anions regulate TORC1, and its ability to bind Fpr1:rapamycin complex in *S. cerevisiae* [47]. According to the authors, elevated levels of superoxide anions modify TORC1 that it is no longer able to fully bind Fpr1:rapamycin complex. To test the possibility that the increased rapamycin-induced autophagy in the *mmi1*Δ strain results from a lower pool of superoxide anions, we stained the WT and the *mmi1*Δ cells with dihydroethidium (DHE), a superoxide anions sensitive probe. The exponentially growing cells were treated with rapamycin for 18 h to reach the post-diauxic growth phase, then the cells were labeled with DHE, and analyzed by FACS flow cytometer. As shown in Figure 6, *mmi1*Δ cells exhibited significantly decreased amount of superoxide anions compared to the WT cells. This suggested that the decreased pool of superoxide anions in *mmi1Δ* cells might contribute to the stronger interaction between rapamycin/Fpr1 and TORC1 and, hence, it might promote the non-selective autophagy. To further test this hypothesis, we analyzed sensitivity of the WT and the *mmi1Δ* strains to rapamycin. We assume that the *mmi1Δ* strain should be more sensitive to rapamycin compared to WT strain if the decreased amount of ROS in *mmi1Δ* strain facilitate binding of Fpr1:rapamycin complex to TORC1. We cultivated the WT strain and the *mmi1Δ* strain in the presence of increasing concentrations of rapamycin, and calculated survival curves as described in [49]. As shown in Figure 7, the WT and the *mmi1Δ* strains exhibited the same sensitivity to rapamycin. These results indicate no correlation between the decreased ROS production and sensitivity to rapamycin in the *mmi1Δ* strain. This suggests that the decreased ROS production in the *mmi1Δ* strain likely does not facilitate the Fpr1:rapamycin binding to TORC1.

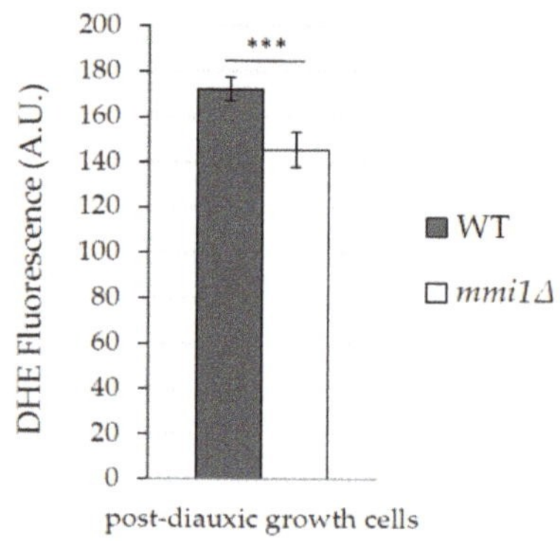

Figure 6. In post diauxic growth phase the *mmi1Δ* strain exhibits lower amount of superoxide anions compared to the WT strain. The amount of superoxide anions was determined in the WT and *mmi1Δ* strains 18 h after rapamycin treatment (200 nM). The production of superoxide anions was measured by dihydroethidium (DHE, 15 μg/mL, 1 h, 30 °C) and samples were analyzed by the FACS flow cytometer. Results are means ± SD of two independent experiments performed in triplicates (n = 6). The statistical evaluation was performed by using two way analysis of variance (ANOVA). *** $p < 0.001$.

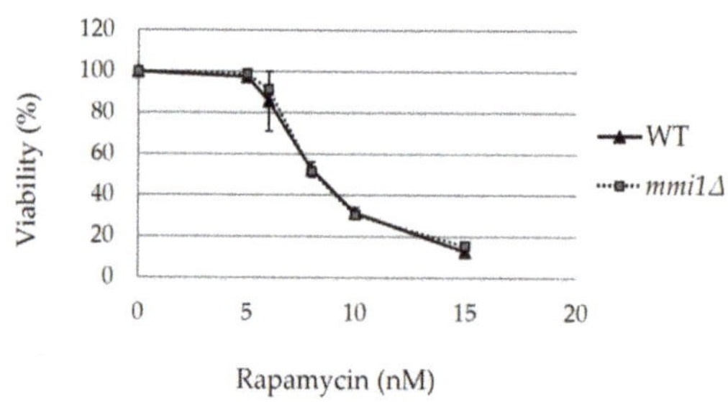

Figure 7. Sensitivity to rapamycin is not influenced in the *mmi1Δ* strain. WT and *mmi1Δ* cells were incubated in the presence of an indicated concentration of rapamycin. Survival curves for the WT and the *mmi1Δ* strains were generated from outgrowth curves and represent the average viabilities of four biological replicates for each strain. Results are means ± SD of two independent experiments performed in duplicates (n = 4).

4. Discussion

As TCTP is a conserved protein from yeast to human and autophagy is a conserved protein degradation pathway, we speculated that examination of the role of yeast TCTP (Mmi1 protein) in autophagy might help us to understand controversial results from higher eukaryotic cells, indicating both positive and negative effects on autophagy [10,11]. Using batch-cultivated yeast cells of *S. cerevisiae*, we demonstrated that in exponentially growing cells Mmi1 protein had no effect on basal or nitrogen starvation-induced bulk non-selective autophagy (Figure 1). However, if exponentially growing cells were treated with rapamycin, Mmi1 negatively affected autophagy (Figures 2–4) when the cells entered the post-diauxic growth phase (Figure 5). Further, in the post-diauxic growth phase we detected lower amount of superoxide radicals in *mmi1Δ* cells compared to WT cells (Figure 6). Our results also indicate that WT and *mmi1Δ* cells possess same sensitivity to rapamycin (Figure 7).

Previously, Chen and colleagues suggested that mammalian TCTP could positively regulate autophagy. By using African green monkey kidney fibroblast-like cells (COS-7 cells) they demonstrated that TCTP knockdown inhibits autophagy under hypoxic or serum starvation conditions [10]. The effect of hypoxic conditions on autophagy has not been tested in this study and awaits further exploration. However, when the exponentially growing yeast cells were shifted from the rich YPD medium to nitrogen starvation conditions, high autophagy levels were induced but no effect of Mmi1 was detected. These results are consistent with the finding that a very slight or no effect on autophagy was previously detected in response to nutrient starvation condition in the knockdown TCTP human HeLa cell line and mouse embryonic fibroblasts (MEFs), haploinsufficient in TCTP expression [11].

Further, contrary to a positive role of TCTP on autophagy, Bae and colleagues reported that TCTP/TPT1 negatively regulates autophagy [11]. They detected increased autophagy in HeLa cells transiently transfected with *TPT1* shRNA or in mouse embryonic fibroblasts (MEFs) from heterozygote knockout mice embryos ($Tpt1^{+/-}$). The negative autophagy regulation was potentiated by rapamycin and an increased autophagy was also demonstrated in vivo in livers and kidneys of *Tpt1* heterozygote mice [11]. Our results indicate that in yeast cells Mmi1 negatively regulates the rapamycin-induced autophagy (Figures 2 and 3) when glucose is exhausted from media (Figure 5B) and the cells enter post-diauxic growth phase. In contrast, no effect on autophagy is detected under nitrogen starvation condition (Figure 1) when glucose is still present in the medium (Figure 5B).

It is generally accepted that both rapamycin treatment and nitrogen starvation inhibit downstream the TOR signaling pathway that results in repression of protein translation and proliferation and leads to autophagy stimulation [31,32]. However, our results indicate a different effect of Mmi1 on autophagy, based on the used conditions. The rapamycin-induced autophagy might differ from the nitrogen starvation-induced autophagy somewhere within TOR inhibition and/or downstream of TOR signaling that happens during post diauxic growth phase that is not present during nitrogen starvation.

Rapamycin has been thought to fully deactivate the budding yeast TORC1 and driving cells into a quiescent/G0 state [33]. However, this dogma has changed since many groups reported only slowed proliferation upon rapamycin treatment [47,53–55] and only partial inhibition of yeast TORC1 [54]. In fact, rapamycin also partially inactivates mammalian mTORC1 [57,58]. It seems evident that maintained proliferative activity upon rapamycin treatment is crucial for detection of increased autophagy induction in *mmi1Δ* cells in our study. Interestingly, we have noticed that upon rapamycin addition to exponentially growing cells both, the WT and the *mmi1Δ* strain grow similarly (Figure 5A), and the *mmi1Δ* strain lost its slow growth phenotype [19]. Since Mmi1 possesses pro-survival activity, we might speculate that during the exponential growth Mmi1 possesses some activity that is directly regulated by TORC1.

Upon rapamycin treatment we detected the increased autophagy in *mmi1Δ* cells of the post-diauxic growth phase. This phenotype could be rescued by the wild type *MMI1* gene inserted into the *mmi1Δ* strain (Figure 4). At the diauxic shift cells switch from fermentation to respiration and from rapid proliferation to slow proliferation [59]. Further, PKA and TOR pathways are downregulated, and PKC and Snf1 pathways are activated, the former transiently [60], the latter important for the induction of a carbon starvation autophagy in cells undergoing the diauxic shift [61]. Importantly, the carbon starvation autophagy is not induced when the cells are grown in glucose medium, and then shifted to carbon starvation media [61]. It requires an absence of catabolite repression [61] that is responsible for a preferential utilization of glucose to other carbon sources and strict repression of respiration [62].

At the moment the mechanism how Mmi1 negatively affects the rapamycin-induced autophagy in the post-diauxic phase is unknown and awaits a detailed exploration.

However, it has been reported that the reactive oxygen species can hamper inhibitory activity of rapamycin in *S. cerevisiae* by oxidative damage to yeast TORC1 [47]. Our results indicate that in the post-diauxic growth phase WT cells exhibit a higher amount of reactive oxidative species compared to *mmi1*Δ cells (Figure 6). Since WT and *mmi1*Δ cells exhibit the same sensitivity to rapamycin (Figure 7), it seems unlikely that the decreased ROS production in the *mmi1*Δ strain influences rapamycin binding to TORC1. Further, the reactive oxygen species accumulation was shown to be critical for autophagy induction during nutrient starvation conditions in mammalian cells [63] and a number of studies indicate autophagy regulation by redox signaling [64]. In *S. cerevisie*, activity of cysteine protease Atg4 could be regulated by the redox state and, hence, it may regulate autophagosome formation [65]. Recently, ethanol stress-induced autophagy was reported to also be regulated by ROS [66].

Here, we used exponentially growing cells and induced autophagy by nitrogen starvation or rapamycin treatment. These approaches were introduced already in 1990s [31,32] and have been used by many groups to trigger autophagy in cells previously grown in the rich YPD medium. Nevertheless, in a natural habitat, cells face large variations in nutrients, and some autophagy roles in cellular metabolism seems to be still unexplored [67]. In this respect, Iwama and Ohsumi recently reported that a bulk autophagy is activated in batch culture on low glucose media based on available carbon sources [68]. Further, Horie and colleagues reported that iron recycling via autophagy is critical for transition from glycolytic to respiratory growth [67]. Several experiments are now in progress to test the effect of Mmi1 on autophagy induced during cell aging and quiescence. So far, our results support the role of Mmi1/TCTP as a negative regulator in the rapamycin-induced non-selective autophagy in eukaryotic cells.

Supplementary Materials: The following are available online at http://www.mdpi.com/2073-4409/9/1/138/s1, Figure S1: Growth of WT and *mmi1*Δ strains in nitrogen starvation media.

Author Contributions: J.V.; designed and performed the experiments, analyzed data, and wrote the manuscript, J.H.; reviewed the manuscript and was responsible for funding acquisition and project administration. All authors have read and agreed to the published version of the manuscript.

Funding: This research was supported by the grant from the Czech Science Foundation CSF16-05497S (J.H.) and by the fellowship from J. W. Fulbright Commission (Prague, Czech Republic) (J.V.).

Acknowledgments: We are very grateful to Suresh Subramani for enabling the learning the autophagy methods in his laboratory at the University of California, San Diego, and for helpful discussions. We also thank to Mark Rinnerthaler (University of Salzburg, Austria) and Ivana Malcova for critical comments on the manuscript, Jaroslav Vojta for the help with statistical evaluation of the data, and Lenka Novakova for her technical assistance.

Conflicts of Interest: The authors declare no conflict of interest.

References

1. Chen, S.H.; Wu, P.-S.; Chou, C.-H.; Yan, Y.-T.; Liu, H.; Weng, S.-Y.; Yang-Yen, H.-F. A Knockout Mouse Approach Reveals that TCTP Functions as an Essential Factor for Cell Proliferation and Survival in a Tissue- or Cell Type–specific Manner. *Mol. Boil. Cell* **2007**, *18*, 2525–2532. [CrossRef] [PubMed]
2. Hsu, Y.-C.; Chern, J.J.; Cai, Y.; Liu, M.; Choi, K.-W. Drosophila TCTP is essential for growth and proliferation through regulation of dRheb GTPase. *Nature* **2007**, *445*, 785–788. [CrossRef] [PubMed]
3. Liu, Z.-L.; Xu, J.; Ling, L.; Zhang, R.; Shang, P.; Huang, Y.-P. CRISPR disruption of TCTP gene impaired normal development in the silkworm *Bombyx mori*. *Insect Sci.* **2018**, *26*, 973–982. [CrossRef]
4. Susini, L.; Besse, S.; Duflaut, D.; Lespagnol, A.; Beekman, C.; Fiucci, G.; Atkinson, A.R.; Busso, D.; Poussin, P.; Marine, J.-C.; et al. TCTP protects from apoptotic cell death by antagonizing bax function. *Cell Death Differ.* **2008**, *15*, 1211–1220. [CrossRef]
5. Rho, S.B.; Lee, J.H.; Park, M.S.; Byun, H.J.; Kang, S.; Seo, S.S.; Kim, J.Y.; Park, S.Y. Anti-Apoptotic Protein Tctp Controls the Stability of the Tumor Suppressor P53. *FEBS Lett.* **2011**, *585*, 29–35. [CrossRef]
6. Li, F.; Zhang, D.; Fujise, K. Characterization of Fortilin, a Novel Antiapoptotic Protein. *J. Boil. Chem.* **2001**, *276*, 47542–47549. [CrossRef]

7. Zhang, D.; Li, F.; Weidner, D.; Mnjoyan, Z.H.; Fujise, K. Physical and Functional Interaction between Myeloid Cell Leukemia 1 Protein (MCL1) and Fortilin. *J. Boil. Chem.* **2002**, *277*, 37430–37438. [CrossRef]
8. Bommer, U.-A. The Translational Controlled Tumour Protein TCTP: Biological Functions and Regulation. *Results Probl. Cell Differ.* **2017**, *64*, 69–126.
9. Thomas, G.; Luther, H. Transcriptional and translational control of cytoplasmic proteins after serum stimulation of quiescent Swiss 3T3 cells. *Proc. Natl. Acad. Sci. USA* **1981**, *78*, 5712–5716. [CrossRef]
10. Chen, K.; Huang, C.; Yuan, J.; Cheng, H.; Zhou, R. Long-Term Artificial Selection Reveals a Role of TCTP in Autophagy in Mammalian Cells. *Mol. Boil. Evol.* **2014**, *31*, 2194–2211. [CrossRef]
11. Bae, S.-Y.; Byun, S.; Bae, S.H.; Min, S.; Woo, H.A.; Lee, K. TPT1 (tumor protein, translationally-controlled 1) negatively regulates autophagy through the BECN1 interactome and an MTORC1-mediated pathway. *Autophagy* **2017**, *13*, 820–833. [CrossRef]
12. Tsukada, M.; Ohsumi, Y. Isolation and characterization of autophagy-defective mutants of Saccharomyces cerevisiae. *FEBS Lett.* **1993**, *333*, 169–174. [CrossRef]
13. Torggler, R.; Papinski, D.; Kraft, C. Assays to Monitor Autophagy in Saccharomyces cerevisiae. *Cells* **2017**, *6*, 23. [CrossRef]
14. Nakatogawa, H.; Suzuki, K.; Kamada, Y.; Ohsumi, Y. Dynamics and diversity in autophagy mechanisms: Lessons from yeast. *Nat. Rev. Mol. Cell Boil.* **2009**, *10*, 458–467. [CrossRef]
15. Farré, J.-C.; Subramani, S. Mechanistic insights into selective autophagy pathways: Lessons from yeast. *Nat. Rev. Mol. Cell Boil.* **2016**, *17*, 537–552. [CrossRef]
16. Galdieri, L.; Mehrotra, S.; Yu, S.; Vancura, A. Transcriptional Regulation in Yeast during Diauxic Shift and Stationary Phase. *OMICS: A J. Integr. Boil.* **2010**, *14*, 629–638. [CrossRef]
17. Swinnen, E.; Wanke, V.; Roosen, J.; Smets, B.; Dubouloz, F.; Pedruzzi, I.; Cameroni, E.; De Virgilio, C.; Winderickx, J. Rim15 and the crossroads of nutrient signalling pathways in Saccharomyces cerevisiae. *Cell Div.* **2006**, *1*, 3. [CrossRef]
18. Fleischer, T.C.; Weaver, C.M.; McAfee, K.J.; Jennings, J.L.; Link, A.J. Systematic identification and functional screens of uncharacterized proteins associated with eukaryotic ribosomal complexes. *Genome Res.* **2006**, *20*, 1294–1307. [CrossRef]
19. Rinnerthaler, M.; Jarolim, S.; Heeren, G.; Palle, E.; Perju, S.; Klinger, H.; Bogengruber, E.; Madeo, F.; Braun, R.J.; Breitenbach-Koller, L.; et al. Mmi1 (Ykl056c, Tma19), the Yeast Orthologue of the Translationally Controlled Tumor Protein (Tctp) Has Apoptotic Functions and Interacts with Both Microtubules and Mitochondria. *Biochim. Et Biophys. Acta* **2006**, *1757*, 631–638. [CrossRef]
20. A Kulak, N.; Pichler, G.; Paron, I.; Nagaraj, N.; Mann, M. Minimal, encapsulated proteomic-sample processing applied to copy-number estimation in eukaryotic cells. *Nat. Methods* **2014**, *11*, 319–324. [CrossRef]
21. Murphy, J.P.; Stepanova, E.; Everley, R.A.; Paulo, J.A.; Gygi, S.P. Comprehensive Temporal Protein Dynamics during the Diauxic Shift in Saccharomyces cerevisiae. *Mol. Cell. Proteom.* **2015**, *14*, 2454–2465. [CrossRef] [PubMed]
22. Delaney, J.R.; Murakami, C.J.; Olsen, B.; Kennedy, B.K.; Kaeberlein, M. Quantitative evidence for early life fitness defects from 32 longevity-associated alleles in yeast. *Cell Cycle* **2011**, *10*, 156–165. [CrossRef] [PubMed]
23. Chong, Y.T.; Koh, J.L.; Friesen, H.; Duffy, S.K.; Cox, M.J.; Moses, A.; Moffat, J.; Boone, C.; Andrews, B.J. Yeast Proteome Dynamics from Single Cell Imaging and Automated Analysis. *Cell* **2015**, *161*, 1413–1424. [CrossRef]
24. Bommer, U.A.; Iadevaia, V.; Chen, J.; Knoch, B.; Engel, M.; Proud, C.G. Growth-Factor Dependent Expression of the Translationally Controlled Tumour Protein Tctp Is Regulated through the Pi3-K/Akt/Mtorc1 Signalling Pathway. *Cell. Signal.* **2015**, *27*, 1557–1568. [CrossRef]
25. Bischof, J.; Salzmann, M.; Streubel, M.K.; Hasek, J.; Geltinger, F.; Duschl, J.; Bresgen, N.; Briza, P.; Haskova, D.; Lejskova, R.; et al. Clearing the outer mitochondrial membrane from harmful proteins via lipid droplets. *Cell Death Discov.* **2017**, *3*, 17016. [CrossRef]
26. Rinnerthaler, M.; Lejskova, R.; Groušl, T.; Stradalova, V.; Heeren, G.; Richter, K.; Breitenbach-Koller, L.; Malínský, J.; Hasek, J.; Breitenbach, M. Mmi1, the Yeast Homologue of Mammalian TCTP, Associates with Stress Granules in Heat-Shocked Cells and Modulates Proteasome Activity. *PLoS ONE* **2013**, *8*, e77791. [CrossRef]
27. Stark, C.; Breitkreutz, B.-J.; Reguly, T.; Boucher, L.; Breitkreutz, A.; Tyers, M. BioGRID: A general repository for interaction datasets. *Nucleic Acids Res.* **2006**, *34*, D535–D539. [CrossRef]

28. Cheong, H.; Klionsky, D.J. Chapter 1 Biochemical Methods to Monitor Autophagy-Related Processes in Yeast. *Methods Enzymol.* **2008**, *451*, 1–26.
29. Wang, X.; Li, S.; Liu, Y.; Ma, C. Redox Regulated Peroxisome Homeostasis. *Redox Biol.* **2015**, 104–108. [CrossRef]
30. Yorimitsu, T.; Klionsky, D.J. Autophagy: Molecular machinery for self-eating. *Cell Death Differ.* **2005**, *12*, 1542–1552. [CrossRef]
31. Takeshige, K.; Baba, M.; Tsuboi, S.; Noda, T.; Ohsumi, Y. Autophagy in yeast demonstrated with proteinase-deficient mutants and conditions for its induction. *J. Cell Boil.* **1992**, *119*, 301–311. [CrossRef]
32. Noda, T. Tor, a Phosphatidylinositol Kinase Homologue, Controls Autophagy in Yeast. *J. Boil. Chem.* **1998**, *273*, 3963–3966. [CrossRef]
33. Barbet, N.C.; Schneider, U.; Helliwell, S.B.; Stansfield, I.; Tuite, M.F.; Hall, M.N. TOR controls translation initiation and early G1 progression in yeast. *Mol. Boil. Cell* **1996**, *7*, 25–42. [CrossRef]
34. Thomas, G.; Hall, M.N. TOR signalling and control of cell growth. *Curr. Opin. Cell Boil.* **1997**, *9*, 782–787. [CrossRef]
35. Parzych, K.R.; Klionsky, D.J. An Overview of Autophagy: Morphology, Mechanism, and Regulation. *Antioxid. Redox Signal.* **2014**, *20*, 460–473. [CrossRef]
36. Ohsumi, Y. Historical Landmarks of Autophagy Research. *Cell Res.* **2014**, *24*, 9–23. [CrossRef] [PubMed]
37. Kawamata, T.; Kamada, Y.; Suzuki, K.; Kuboshima, N.; Akimatsu, H.; Ota, S.; Ohsumi, M.; Ohsumi, Y. Characterization of a novel autophagy-specific gene, ATG29. *Biochem. Biophys. Res. Commun.* **2005**, *338*, 1884–1889. [CrossRef] [PubMed]
38. Kiel, J.A.K.W. Autophagy in unicellular eukaryotes. *Philos. Trans. R. Soc. B: Boil. Sci.* **2010**, *365*, 819–830. [CrossRef] [PubMed]
39. Liu, E.Y.; Ryan, K.M. Autophagy and cancer-issues we need to digest. *J. Cell Sci.* **2012**, *125*, 2349–2358. [CrossRef]
40. Brachmann, C.B.; Davies, A.; Cost, G.J.; Caputo, E.; Li, J.; Hieter, P.; Boeke, J.D. Designer deletion strains derived fromSaccharomyces cerevisiae S288C: A useful set of strains and plasmids for PCR-mediated gene disruption and other applications. *Yeast* **1998**, *14*, 115–132. [CrossRef]
41. Noda, T.; Matsuura, A.; Wada, Y.; Ohsumi, Y. Novel System for Monitoring Autophagy in the Yeast Saccharomyces cerevisiae. *Biochem. Biophys. Res. Commun.* **1995**, *210*, 126–132. [CrossRef]
42. Bicknell, A.A.; Tourtellotte, J.; Niwa, M. Late Phase of the Endoplasmic Reticulum Stress Response Pathway Is Regulated by Hog1 MAP Kinase*. *J. Boil. Chem.* **2010**, *285*, 17545–17555. [CrossRef]
43. Noda, T.; Klionsky, D.J. The Quantitative Pho8delta60 Assay of Nonspecific Autophagy. *Methods Enzymol.* **2018**, *451*, 33–42.
44. Kramer, M.H.; Farré, J.-C.; Mitra, K.; Yu, M.K.; Ono, K.; Demchak, B.; Licon, K.; Flagg, M.; Balakrishnan, R.; Cherry, J.M.; et al. Active Interaction Mapping Reveals the Hierarchical Organization of Autophagy. *Mol. Cell* **2017**, *65*, 761–774. [CrossRef]
45. Suzuki, K.; Kirisako, T.; Kamada, Y.; Mizushima, N.; Noda, T.; Ohsumi, Y. The pre-autophagosomal structure organized by concerted functions of APG genes is essential for autophagosome formation. *EMBO J.* **2001**, *20*, 5971–5981. [CrossRef]
46. Rueden, C.T.; Schindelin, J.; Hiner, M.C.; DeZonia, B.E.; Walter, A.E.; Arena, E.T.; Eliceiri, K.W. Imagej2: Imagej for the Next Generation of Scientific Image Data. *BMC Bioinform.* **2017**, *18*, 529. [CrossRef]
47. Neklesa, T.K.; Davis, R.W. Superoxide anions regulate TORC1 and its ability to bind Fpr1:rapamycin complex. *Proc. Natl. Acad. Sci. USA* **2008**, *105*, 15166–15171. [CrossRef]
48. Robinson, K.M.; Janes, M.S.; Pehar, M.; Monette, J.S.; Ross, M.F.; Hagen, T.M.; Murphy, M.P.; Beckman, J.S. Selective fluorescent imaging of superoxide in vivo using ethidium-based probes. *Proc. Natl. Acad. Sci. USA* **2006**, *103*, 15038–15043. [CrossRef]
49. Murakami, C.; Kaeberlein, M. Quantifying Yeast Chronological Life Span by Outgrowth of Aged Cells. *J. Vis. Exp.* **2009**. [CrossRef]
50. Kirisako, T.; Ichimura, Y.; Okada, H.; Kabeya, Y.; Mizushima, N.; Yoshimori, T.; Ohsumi, M.; Takao, T.; Noda, T.; Ohsumi, Y. The Reversible Modification Regulates the Membrane-Binding State of Apg8/Aut7 Essential for Autophagy and the Cytoplasm to Vacuole Targeting Pathway. *J. Cell Boil.* **2000**, *151*, 263–276. [CrossRef]

51. Matsuura, A.; Tsukada, M.; Wada, Y.; Ohsumi, Y. Apg1p, a novel protein kinase required for the autophagic process in Saccharomyces cerevisiae. *Gene* **1997**, *192*, 245–250. [CrossRef]
52. An, Z.; Tassa, A.; Thomas, C.; Zhong, R.; Xiao, G.; Fotedar, R.; Tu, B.P.; Klionsky, D.J.; Levine, B. Autophagy Is Required for G(1)/G(0) Quiescence in Response to Nitrogen Starvation in Saccharomyces Cerevisiae. *Autophagy* **2014**, *10*, 1702–1711. [CrossRef]
53. Rallis, C.; Codlin, S.; Bahler, J. TORC1 signaling inhibition by rapamycin and caffeine affect lifespan, global gene expression, and cell proliferation of fission yeast. *Aging Cell* **2013**, *12*, 563–573. [CrossRef]
54. Evans, S.K.; Burgess, K.E.; Gray, J.V. Recovery from Rapamycin: Drug-Insensitive Activity of Yeast Target of Rapamycin Complex 1 (Torc1) Supports Residual Proliferation That Dilutes Rapamycin among Progeny Cells. *J. Biol. Chem.* **2014**, *289*, 26554–26565. [CrossRef]
55. Dikicioglu, D.; Eke, E.D.; Eraslan, S.; Oliver, S.G.; Kirdar, B. Saccharomyces cerevisiae adapted to grow in the presence of low-dose rapamycin exhibit altered amino acid metabolism. *Cell Commun. Signal.* **2018**, *16*, 85. [CrossRef] [PubMed]
56. Morgan, J.T.; Fink, G.R.; Bartel, D.P. Excised linear introns regulate growth in yeast. *Nature* **2019**, *565*, 606–611. [CrossRef]
57. Thoreen, C.C.; Kang, S.A.; Chang, J.W.; Liu, Q.; Zhang, J.; Gao, Y.; Reichling, L.J.; Sim, T.; Sabatini, D.M.; Gray, N.S. An ATP-competitive Mammalian Target of Rapamycin Inhibitor Reveals Rapamycin-resistant Functions of mTORC1. *J. Boil. Chem.* **2009**, *284*, 8023–8032. [CrossRef]
58. Feldman, M.E.; Apsel, B.; Uotila, A.; Loewith, R.; Knight, Z.A.; Ruggero, D.; Shokat, K.M. Active-site inhibitors of mTOR target rapamycin-resistant outputs of mTORC1 and mTORC2. *PLoS Biol.* **2009**, *7*, e1000038. [CrossRef]
59. Gray, J.V.; Petsko, G.A.; Johnston, G.C.; Ringe, D.; Singer, R.A.; Werner-Washburne, M. "Sleeping Beauty": Quiescence in Saccharomyces cerevisiae†. *Microbiol. Mol. Boil. Rev.* **2004**, *68*, 187–206. [CrossRef]
60. Krause, S.A.; Gray, J.V. The protein kinase C pathway is required for viability in quiescence in Saccharomyces cerevisiae. *Curr. Boil.* **2002**, *12*, 588–593. [CrossRef]
61. Adachi, A.; Koizumi, M.; Ohsumi, Y. Autophagy induction under carbon starvation conditions is negatively regulated by carbon catabolite repression. *J. Boil. Chem.* **2017**, *292*, 19905–19918. [CrossRef] [PubMed]
62. Kayikci, Ö.; Nielsen, J. Glucose repression in Saccharomyces cerevisiae. *FEMS Yeast Res.* **2015**, *15*.
63. Scherz-Shouval, R.; Shvets, E.; Fass, E.; Shorer, H.; Gil, L.; Elazar, Z. Reactive oxygen species are essential for autophagy and specifically regulate the activity of Atg4. *EMBO J.* **2007**, *26*, 1749–1760. [CrossRef] [PubMed]
64. Lee, J.; Giordano, S.; Zhang, J. Autophagy, Mitochondria and Oxidative Stress: Cross-Talk and Redox Signalling. *Biochem. J.* **2011**, *441*, 523–540. [CrossRef] [PubMed]
65. Pérez-Pérez, M.E.; Zaffagnini, M.; Marchand, C.H.; Crespo, J.L.; Lemaire, S.D. The yeast autophagy protease Atg4 is regulated by thioredoxin. *Autophagy* **2014**, *10*, 1953–1964. [CrossRef] [PubMed]
66. Jing, H.; Liu, H.; Zhang, L.; Gao, J.; Song, H.; Tan, X. Ethanol induces autophagy regulated by mitochondrial ROS in Saccharomyces cerevisiae. *J. Microbiol. Biotechnol.* **2018**, *28*, 1982–1991. [CrossRef]
67. Horie, T.; Kawamata, T.; Matsunami, M.; Ohsumi, Y. Recycling of iron via autophagy is critical for the transition from glycolytic to respiratory growth. *J. Boil. Chem.* **2017**, *292*, 8533–8543. [CrossRef]
68. Iwama, R.; Ohsumi, Y. Analysis of autophagy activated during changes in carbon source availability in yeast cells. *J. Boil. Chem.* **2019**, *294*, 5590–5603. [CrossRef]

cells

MDPI

Review

Dysregulation of TCTP in Biological Processes and Diseases

Ulrich-Axel Bommer [1,*] and Adam Telerman [2,*]

1 School of Medicine, Faculty of Science, Medicine & Health, University of Wollongong, Wollongong, NSW 2522, Australia
2 Institut Gustave Roussy, Unite Inserm U981, 94805 Villejuif, France
* Correspondence: ubommer@uow.edu.au (U.-A.B.); atelerman@gmail.com (A.T.)

Received: 19 June 2020; Accepted: 3 July 2020; Published: 7 July 2020

Abstract: Translationally controlled tumor protein (TCTP), also called histamine releasing factor (HRF) or fortilin, is a multifunctional protein present in almost all eukaryotic organisms. TCTP is involved in a range of basic cell biological processes, such as promotion of growth and development, or cellular defense in response to biological stresses. Cellular TCTP levels are highly regulated in response to a variety of physiological signals, and regulatory mechanism at various levels have been elucidated. Given the importance of TCTP in maintaining cellular homeostasis, it is not surprising that dysregulation of this protein is associated with a range of disease processes. Here, we review recent progress that has been made in the characterisation of the basic biological functions of TCTP, in the description of mechanisms involved in regulating its cellular levels and in the understanding of dysregulation of TCTP, as it occurs in disease processes such as cancer.

Keywords: TCTP (HRF; fortilin); growth and development; biological stress reactions; autophagy; regulation of protein synthesis; regulated protein degradation; cancer; cardiovascular diseases

1. General Overview

TCTP (also referred to as: HRF, fortilin, P23; gene symbol tpt-1; in yeast: TMA19 or Mmi1) was first described almost 40 years ago as a growth factor-induced, translationally controlled protein in murine cell lines [1,2]. The elucidation of its biological importance was initially hampered by the fact that, based on its amino acid sequence, it is not related to any other known protein family. It took a considerable international research effort, spread over more than three decades, to piece together the puzzle and reveal our current picture of the biological function of TCTP and of its involvement in various diseases. Several review articles on TCTP/HRF/fortilin have been published previously, and the first 'TCTP book' appeared in 2017, a compilation of 16 individual review articles covering a wide range of topics relating to this multifunctional protein [3]. Since then, a considerable number of new papers reported various novel aspects on TCTP, and it is the aim of this review to summarise these more recent additions to the 'TCTP story'. Being part of the Special Issue 'Role of TCTP in Cell Biological and Disease Processes' in *Cells*, our review will only briefly touch on those topics that are the subject of the other review articles within this Special Issue. However, we will discuss the research papers of the Special Issue, where appropriate, to put these into context.

2. Importance of TCTP in Core Cell Biological and Stress Reactions

The involvement of TCTP in a multitude of cell biological processes is reflected by the large number of interacting proteins. Compilations of the many interaction partners of TCTP have been published in the above-mentioned book [4–6]. A more recent study, using current proteomics approaches combined with in silico network and gene ontology enrichment analysis, identified 113 interacting proteins [7], which were allocated to the following gene network clusters: translation, stress response, cytoskeleton, signal transduction, proteasomes, nucleosomes and transcription, RNA binding and metabolic enzymes. These clusters can be roughly associated with the following three core biological processes. We will discuss recent advances on the participation of TCTP in these processes, as summarised in Table 1 below.

2.1. Growth and Developmental Processes

There is a considerable body of evidence for the involvement of TCTP in cell and organ growth, as well as in developmental processes, most aspects of which were reviewed in the following chapters of the 'TCTP book': cell cycle progression [5,8], early development [5,9], and organ growth and development [9–11]. Since then, several new studies have extended our insight into potential mechanisms, through which TCTP might participate in these processes.

Regarding cell cycle regulation in early development, Jeon et al. showed that TCTP regulates spindle assembly during postovulatory aging of mouse oocytes, thereby preventing deterioration of oocyte quality [12]. The targeting of yet another mechanism for cell cycle control by TCTP was revealed in a more recent paper [13]. The authors showed that TCTP interacts with CSN4, a subunit of the COP9 signalosome complex, which controls the G1/S transition of the cell cycle through regulating Cullin–Ring ubiquitin ligases. This mechanism is conserved between plants (*Arabidopsis*) and insects (*Drosophila*). The latter observation is consistent with another report showing that disruption of TCTP using CRISPR/Cas9 in the silkworm *Bombyx mori* resulted in developmental arrest and in subsequent lethality in the third instar larvae, caused by defects in proliferation of intestinal epithelial cells [14]. Similarly, a recent report demonstrated the importance of TCTP in stem cells of the midgut in *Drosophila* for tissue homeostasis and regeneration [15]. As potential mechanisms, the authors suggest the regulation of protein synthesis by TCTP (Section 2.2. below) and crosstalk with two important growth regulating signalling pathways.

The importance of TCTP for organ development was supported by a new report showing that TCTP promotes liver regeneration via mTORC2/Akt signalling [16], by studies on axon guidance [17–19] and on brain development [20]. Roque et al. demonstrated the role of TCTP for axon guidance and development in the visual system of *Xenopus laevis* [18] and the importance of its localised translational regulation in axonal growth cones [17]. A recent paper confirmed the importance of TCTP for general brain development in mice [20]. Conditional TCTP-knockout mice displayed retardation in brain development and died at the perinatal stage. An interesting case of the involvement of TCTP in organ development was reported for plants [21]. The authors demonstrated that in *Arabidopsis thaliana*, the mRNA (and protein) of the AtTCTP1 protein is transported over a long distance from scions into the roots, where it provides a signal to govern the formation of lateral roots. In contrast, the 'endogenous' AtTCTP1, locally produced in the root cells, drives the elongation of the roots. Thus, TCTP is a molecule crucial for establishing the root architecture of plants.

2.2. Regulation of Protein Synthesis and Degradation

At the beginning of this century, the association of TCTP with ribosomes [22] and the translational machinery [23] was identified in yeast. A more specific study demonstrated the interaction of TCTP with translation elongation factor eEF1A and its guanine nucleotide exchange factor (GEF), eEF1B [24]. Subsequently, the interaction of TCTP with elongation factors of the EF1 family and with other components of the translational apparatus was confirmed both in human cells [7,25] and in *Drosophila* [15]. Consistent with this is the observation that the genes of ribosomal proteins, elongation factors, and of TCTP all belong to the class of 'TOP genes', whose mRNAs have a common signature, the 5′-terminal oligo-pyrimidine tract (5′-TOP) [26], and are therefore translationally regulated [27].

Later, the interaction of TCTP with eEF1B was studied in more detail, using a range of structural methods [28]. This paper demonstrated that TCTP binds to the central acidic region of eEF1B and that in both these proteins the mutually interacting regions are highly conserved in evolution, thus representing the most conserved interaction of TCTP. This conclusion was also supported by the recently published solution structure of TCTP from a unicellular micro-alga [29].

The functional importance of TCTP in its interaction with translation elongation factor 1 and/or its GEFs still needs to be fully clarified. The initial observation was that, through binding to eEF1Bbeta, TCTP impaired the GDP-GTP exchange reaction on eEF1A, stabilising it in its GDP-bound form [24]. A variation of this mechanism was proposed very recently by a Japanese group interested in mechanisms involved in formation of neurofibromatosis type 1 (NF1)-associated tumors. Their work confirmed the interaction of TCTP with EF1A, as well as with its GEF complex consisting of EF1B, EF1G, and EF1D [7]. In addition, this paper showed that the interaction of TCTP with EF1A2, an isoform of EF1A preferentially expressed in neuronal tissue and skeletal muscle, is much stronger compared to that with the normal isoform, EF1A1. The authors concluded that, in NF1-associated tumor cells, TCTP binds to the GDP-bound form of EF1A2, thus preventing its dimerisation and inactivation. In this way, TCTP facilitates the binding of the GEF complex to GDP-bound EF1A2, promoting the GDP-GTP exchange reaction and recycling of EF1A2 [7].

Whether the net effect of TCTP on translation elongation is positive [7] or negative [24], both should result in a general regulation of protein synthesis. However, another well-documented example of translational control that targets the elongation cycle of protein synthesis has been shown to result in a selective translational advantage for specific mRNAs. Elongation factor 2 (EF2)-kinase phosphorylates EF2, thereby slowing down protein synthesis. This results in an increased expression of proteins implicated in cell migration and cancer cell metastasis [30]. It remains to be seen whether the effect of TCTP on EF1 activity may also lead to a preferential translation of certain mRNAs.

Two additional observations confirm the involvement of TCTP in the translational machinery, i.e., the recently reported interaction of TCTP with the receptor of activated protein kinase C (RACK1) [31] and the identification of TCTP as an mRNA-binding protein in HeLa cells (Supplementary Table S1 in [32]). These two observations may be related to each other, since RACK1 was identified as a ribosomal protein that is located close to the ribosomal mRNA entry site [33]. RACK1 serves as a ribosomal scaffolding protein that is involved in targeting the ribosome to various signalling modules and also to specific subcellular locations, such as focal adhesion sites [33].

Several examples of an involvement of TCTP in the regulated degradation of specific proteins have been reported. A review on this topic was published in the TCTP book [34]; here we will briefly summarise the relevant examples (see also Table 1 below):

1. TCTP stabilises the following client proteins by masking their ubiquitination sites, thereby preventing their proteasomal degradation. These are the anti-apoptotic protein Mcl-1 [35], the protein kinase Pim-3, a proto-oncogene [36], the antioxidant enzyme peroxiredoxin PRX1 [37], and the tobacco histidine kinase 1, NTHK1 [38]. This kinase represents the receptor for the phytohormone ethylene in plants. In this way, TCTP reduces the plant response to ethylene, which enhances plant growth and cell proliferation [38]. A specific case is the stabilisation of hypoxia-inducible factor HIF1α, where TCTP does not bind to HIF1α itself, but to the von-Hippel-Landau protein (VHL), which normally acts as E3-ligase for HIF1α. TCTP binding results in ubiquitination of VHL and in its own proteasomal degradation, preventing the degradation of HIF1α [39].

2. For other proteins, TCTP actively promotes degradation. For example, to induce degradation of the tumor suppressor protein p53, TCTP binds to p53-MDM2 complexes and inhibits MDM2 auto-ubiquitination, resulting in MDM2-mediated ubiquitination and degradation of p53 [40]. TCTP also promotes the degradation of cell cycle protein phosphatase Cdc25, which is important for an orderly mitotic exit. TCTP overexpression, as it occurs in hepatocellular carcinoma (HCC), induces ubiquitination and degradation of Cdc25, eventually resulting in chromosome missegregation [41].

3. Apart from regulating the degradation of specific proteins, in yeast, TCTP (TMA19; Mmi1) has also been shown to interact with proteasomal proteins [42] and to colocalise with the proteasome under heat stress conditions [43]. Yeast Mmi1 slightly inhibited proteasome activity, and was found to colocalise with two proteins of a de-ubiquitination complex in heat-stressed cells [43].

2.3. Biological Stress Reactions and Autophagy

The role of TCTP as an anti-apoptotic protein involved in protecting cells in a range of stress conditions (inclusive of DNA damage [44]) in mammalian cells [5,45], in plants [9], and in insects [10] has been reviewed in several articles of the TCTP book [3]. Here, we will just summarise the more recent contributions in this area.

Two recent papers demonstrated the importance of TCTP as a survival factor in mammalian organs. Cai et al. showed that TCTP plays a critical role for the survival of cardiomyocytes and has a protective function against drug-induced cardiac dysfunction in mice [46]. Another paper, published in this Special Issue in *Cells*, studied the importance of TCTP for the development of the central nervous system [20]. The authors generated mice that are disrupted in TCTP expression in neuronal and glial progenitor cells. These mice die at the perinatal stage, and they show slight abnormalities in early brain development that are associated with increased apoptosis, demonstrating that TCTP is a critical protein for cell survival during early neuronal and glial differentiation.

It was well established for quite some time that TCTP is involved in protecting cells in a wide range of stress conditions. However, up until recently, it was not known that it also plays a role in the ER-stress defense program, the unfolded protein response (UPR). This gap has been closed by a recent paper by Pinkaew et al. [47]. IRE1α is one of three main players in initiating the UPR; it has protein kinase and endonuclease activity and is ultimately responsible for the induction of apoptosis, once the protein overload in the ER becomes overwhelming. The authors showed that TCTP (fortilin) is able to bind to phosphorylated IRE1α, thereby preventing the activation of the JNK apoptosis pathway by IRE1α.

The importance of TCTP in biological stress reactions was also demonstrated in non-mammalian systems. De Carvalho et al. overexpressed TCTP from tomatoes in tobacco plants, which resulted in an increased growth rate and an improved performance under salt and osmotic stress conditions [48]. The transgenic plants displayed an increased expression of genes involved in photosynthesis, fatty acid metabolism, and water transport, and this was paralleled by an increased photosynthetic rate. On the other hand, genes involved in ethylene biosynthesis (a plant hormone) were down-regulated by TCTP overexpression. This is consistent with the observation that TCTP binds to the ethylene receptor in tobacco plants and reduces its response to ethylene [38]. Studying the role of TCTP in Trypanosomes, Jojic and co-workers reported that, in the bloodstream form of the parasite, depletion of TCTP expression resulted in a reduced growth rate and also in a slower recovery after heat stress [49]. The importance of TCTP in the development of resistance against the insecticide deltamethrin was investigated in Drosophila kc cells [50]. RNAi-knockdown and overexpression experiments confirmed that TCTP partially protects these cells against deltamethrin-induced cell death.

Over the past decade, autophagy has been recognised as a central biological process, essential for maintaining cellular homeostasis; but the involvement of TCTP in this pathway has only recently been investigated. In the 'TCTP book', this aspect of TCTP function has only briefly been touched on in two articles [5,34] and also here, we will not cover it in much detail, since another review article in this Special Issue discussed the role of TCTP and autophagy in relation to tumorigenesis [51].

So far, the effect of TCTP on autophagy has only been studied in three original reports (Table 1), with partially conflicting results. The earliest of these papers found that TCTP is one of five genes upregulated by artificial selection in the ovaries of domesticated vs. wild pigs [52]. The authors also observed that TCTP is located in the cytoplasm, in a pattern similar to the autophagy protein LC3. In COS-7 cells kept under normoxic conditions, TCTP knockdown resulted in enhanced AMPK activity and an increase in the levels of the LC3-II protein, whereas the opposite was true under hypoxic conditions. The observed effects could be reversed by re-introducing the TCTP gene into the cells via a lentivirus construct. TCTP was also found to interact with the ATG16 complex of autophagic proteins. The authors concluded that TCTP positively regulates autophagy via the AMPK/mTORC1 pathway. Another, more recent paper arrived at a different conclusion i.e., that in HeLa cells, TCTP inhibits autophagy through two separate pathways, 1. the AMPK/mTORC1 signalling pathway by inhibiting

AMPK and activating mTORC1, and 2. through activation of Bcl-2, which in turn leads to an inhibition of the formation of the autophagic Beclin1 complex [53]. Both these studies demonstrated that TCTP is indeed able to modulate autophagy, however, the actual outcome is very much dependent on the precise cell physiologic conditions. Consistent with this, another paper on TCTP regulation of autophagy, published in this Special Issue, observed that TCTP negatively affects rapamycin-induced autophagy in the post-diauxic growth phase in yeast, but not autophagy induced by nitrogen starvation [54].

Table 1. The importance of TCTP in basic biological processes.

Process	Examples	Potential Mechanism	References
Cell growth and organ development	Prevention of oocyte aging Cell cycle progression Insect development Tissue regeneration Brain development Branching of plant roots	Regulates spindle assembly Signalosome interaction Proliferation of epithelial cells Stem cells; protein synthesis (Axon guidance) Initiation of lateral roots	[12] [13] [14] [15,16] [18–20] [21]
Regulation of protein synthesis	Association with proteins of the translational apparatus - TCTP binding to EF1B - TCTP binding to EF1A2 - Interaction with RACK1 - Binding to mRNAs	(in yeast) (in human cells and *Drosophila*) Regulation of the GEF activity of EF1B on EF1A (Part of the mRNA interactome)	[22,23] [7,15] [24,25,28] [7] [31] [32]
Regulation of protein degradation	Stabilisation of: HIF1α, Mcl-1, PIM-3, PRX1, NTHK1 Destabilisation of p53, Cdc25 Binding to proteasomes	Prevention of ubiquitination Induction of ubiquitination Inhibition of proteasomes	[34,39] [35–38] [40,41] [42,43]
Biological stress responses	Cardiomyocyte protection ER stress; (binding to IRE1α) Osmotic stress in plants Heat stress in Trypanosomes	Blocking apoptosis by Bnip3 Blocking apoptosis by JNK Increases photosynthesis	[46] [47] [48] [49]
Autophagy	Stimulation of autophagy Reduction of autophagy Reduction of autophagy	AMPK/mTORC1 pathway AMPK pathway; Beclin1 Rapamycin-induced autophagy	[52] [53] [54]

2.4. *Extracellular Functions of TCTP(HRF)*

Since the discovery of the histamine releasing factor (HRF) activity of TCTP in 1995 [55], it is established that, apart from its many intracellular functions, TCTP also acts as an extracellular molecule, typically in the context of immune reactions associated with allergic diseases. However, we will not expand on this, since the review by Kawakami et al. in this Special Issue [56] provided an up-to-date account on HRF and the current understanding of its involvement in various disease settings and its potential as a new therapeutic target. For a historic overview on the 'HRF story', the reader is referred to an earlier review article by Susan MacDonald [57].

3. Mechanisms of Regulation of Cellular TCTP Protein Levels

Being an anti-apoptotic protein involved in cellular stress responses, it is not surprising that the levels of TCTP are highly regulated. In our previous review article [5], we compiled a list of cell physiologic conditions that resulted in alterations of cellular TCTP levels. We also provided an overview on the principal mechanisms involved in TCTP regulation. As the name of the protein suggests, translational regulation plays an important role among these processes; it represents the most effective means for rapid de novo-synthesis. However, other modes of protein regulation, such as transcriptional control or regulated protein degradation, are also involved (Table 2, below). Since the publication of our review article [5], several new examples and mechanisms of regulation of TCTP levels have been revealed, and here we will largely focus on these more recent developments in this area.

3.1. Transcriptional Regulation

To date, numerous instances of transcriptional regulation of TCTP synthesis have been reported [5], but precise mechanistic investigations were only performed in a few cases. Thiele and coworkers demonstrated that TCTP synthesis is induced by phorbolester (PMA) through the cAMP-PKA signalling pathway and activation of transcription factor CREB [58]. They also studied the regulation of TCTP expression under stress conditions, such as heavy metals, and showed that copper induced TCTP synthesis at both the transcriptional and the translational level [59]. Several similar studies showed that TCTP is transcriptionally regulated under a variety of stress conditions [5].

Of particular relevance for cancer research was the demonstration that the tumor suppressor protein p53, acting as a transcription factor, negatively affects TCTP expression [40]. Since p53 is frequently mutated in cancers, this contributes to the observed deregulation of TCTP synthesis in cancers. A more recent paper reported another example of transcriptional regulation of TCTP expression [31]. The authors studied a specific transcription factor, insulin-response element binding protein 1 (IRE-BP1), whose decreased expression in insulin target tissues contributes to the development of insulin-resistant diabetes in rats. Normally, insulin-induced PI3K signalling results in the proteolytic cleavage of IRE-BP1 and release of the active C-terminal fragment of the protein. In order to study the target genes of IRE-BP1, the investigators analysed the proteome of pancreatic islets from transgenic mice overexpressing the active fragment of IRE-BP1, compared to the corresponding proteome from control mice. The overexpression of two of the identified target genes of IRE-BP1 in the transgenic mice was further confirmed by Western blotting, specifically the genes for RACK1 and for TCTP. Moreover, the authors discovered the interaction of RACK1 with TCTP by Co-IP experiments [31].

3.2. Translational Regulation

TCTP was originally discovered by the fact that its synthesis is induced after serum-stimulation of murine cells by translational activation of its preformed mRNA [1]. By 2016, several additional examples of translational regulation of TCTP synthesis were reported (reviewed in [5]), and among these, two principal translational control mechanism were found to be involved: (1) the growth factor-related induction of TCTP synthesis via the mTORC1-eIF4E signalling pathway [60,61], and (2) the negative regulation of TCTP synthesis via activation of PKR and eIF2α-phosphorylation. The latter occurs under serum-starvation [62] and also under Ca^{2+}-stress conditions [63]. Our original observation that TCTP mRNA is a highly structured molecule that binds to and activates PKR [62] was further confirmed by a recent structural analysis of this mRNA under a various conditions, inclusive of the influence of riboSNitches [64].

Similarly, the regulation of TCTP synthesis through mTORC1 was recently observed in a very specific biological system, the retinal axon. Based on the findings from genome-wide screens, that TCTP mRNA is one of the most abundant mRNAs localised in the growth cone of axons, Roque et al. studied the role of TCTP in the development of the retinal axon in *Xenopus laevis*. They showed that it is indeed a protein essential for the normal development of the visual circuitry in the frog [18]. In a follow-up paper, the authors specifically investigated the translational regulation of TCTP and its importance for axon guidance [17]. They demonstrated that in axons of retinal ganglion cells, local TCTP synthesis is regulated by two axon guidance cues in an mTORC1-dependent fashion. In a completely different biological system, the intestinal stem cells in the midgut of *Drosophila*, yet another example of posttranscriptional regulation of TCTP synthesis was observed [15]. In this case, the Hippo signalling pathway has been implicated. However, the precise mechanism still awaits further characterisation.

The two translational control mechanisms referred to so far that target TCTP mRNA, act on the 5′-TOP (the mTORC1 signalling pathway), or on the highly structured area of the mRNA (for PKR activation). The latter most likely comprises the CG-rich 5′-UTR and a 5′-terminal stretch of the coding region [64]. Other translational control modes, targeting the 3′-UTR of TCTP mRNA, emerged only recently. This is surprising, since for 20 years, we have known that in many organisms, TCTP mRNA occurs in two isoforms, which differ in the length of their 3′-UTRs [65], with the shorter isoform usually being the most abundant one. Therefore, the biological importance of the second isoform is still elusive. In addition, for the regulation of TCTP expression in the axonal growth cone, it was shown that it is the shorter isoform, which is subject to translational regulation [17]. The first example of regulation of TCTP expression via the 3′-UTR of its mRNA was revealed in a recent, very interesting report [66]. The authors studied the expression of TCTP in *Trypanosoma brucei* and found that this unicellular parasite during its life cycle differentially expresses two TCTP paralogs, TCTP1 in the procyclic life form (in the tsetse fly) and TCTP2 in the blood stream form in humans. The two mRNAs largely differ in their 3′-UTRs, with TCTP2-mRNA bearing a 1.5-fold longer one. The authors demonstrated that the 3′-UTRs confer differential stability to these TCTP mRNA isoforms and that, for TCTP2 mRNA, the *cis*-acting element largely resides in the first 160 nucleotides of the 3′-UTR. However, the precise characterisation of the *cis*-acting element awaits further investigation.

Table 2. Mechanisms involved in regulating cellular TCTP levels.

Type of Regulation	Example	Mechanism Involved	References
Transcriptional regulation	Induction by PMA	Transcription factor CREB	[58]
	Copper stress reaction	Transcription and Translation	[59]
	Tumor suppressor p53	Inhibition of tpt1 transcription	[40]
	Insulin response	Transcription factor IRE-BP1	[31]
Translational regulation	Induction of TCTP synthesis	via mTORC1-eIF4E signalling	
	- by growth stimuli	- in muscle	[60] [61]
	- during axon guidance	- in retinal ganglion cells	[17]
	Negative regulation in cell stress responses	Activation of PKR and phosphorylation of eIF2α	[62–64]
	Developmental regulation in Trypanosomes	mRNA stability regulation via the 3′-UTR	[66]
	Regulation by micro-RNAs	miR-130a	[67]
	- in various cancers	miR-27b; miR-145-5p; miR-125-3p	[68–70]
Regulation of protein degradation	Stabilisation of TCTP	- by Mcl-1	[71]
		- by Hsp27 (in prostate cancer)	[72,73]
	DHA destabilises TCTP	Ubiquitin/proteasome pathway	[74]
	Partial degradation in mitosis	Ubiquitin/proteasome pathway	[75]
	Chaperone-mediated autophagy	Lysosomal degradation	[76]

Another mode of posttranscriptional regulation that targets individual mRNAs, typically via their 3′-UTRs, is the regulation by micro-RNAs. These short RNA molecules recruit 'their' target mRNAs to the RNA-induced silencing complex (RISC), which leads to translational arrest or even degradation of the mRNA. However, information about validated cases of TCTP mRNAs being regulated in this way is scarce (Table 2). In 2012, two papers reported that TCTP mRNA is regulated by miRNAs, i.e., miR-130a [67] and miR-27b [68], respectively. In the latter case, the authors observed that miR-27b levels were significantly reduced in tumor tissue of oral cancer patients, resulting in an increase of TCTP protein expression. Similarly, two recent papers reported on the deregulated TCTP expression in cancer, due to low levels of certain miRNAs, i.e., miR-145-5p in prolactinoma [69] and miR-125-3p in lung cancer [70]. In the latter paper, the authors studied the induction of TCTP in lung cells by tobacco smoke carcinogens. They found that miR-125-3p, is able to prevent the expression of a TCTP 3′-UTR reporter gene construct and that this miRNA is significantly down-regulated in xenografts generated from cells pretreated with these carcinogens.

3.3. *Regulation of Protein Degradation*

The mechanisms of regulated degradation of the TCTP protein were reviewed in detail in [34] and were also covered in a review article in this Special Issue in *Cells* [51]. Here, we will just briefly summarise the few cases known today (Table 2). Earlier papers reported the stabilisation of the TCTP (fortilin) protein by the anti-apoptotic protein Mcl-1 [71] and by the small heat shock protein Hsp27 [72], thereby preventing TCTP degradation in specific conditions, such as prostate cancer [73]. It was also shown that the antimalarial drug dihydroartemisinin (DHA), which is also used as an anti-cancer agent, binds TCTP and targets it for the ubiquitin-proteasome degradation pathway [74]. Another paper reported at the same time that, during mitosis and during meiotic exit, TCTP is partially degraded [75]. Our recent study characterised the pathway for specific TCTP degradation as acetylation-dependent chaperone-mediated autophagy (CMA), which eventually leads to lysosomal degradation of the protein and involves the proteins Hsc70 and LAMP-2A [76].

4. Disease Processes Involving Dysregulation of TCTP

4.1. Mechanisms of Cancer Promotion by TCTP

When we initially discovered that TCTP was overexpressed in a majority of tumors [77], it became clear that its inhibition could potentially result in decreased tumorigenicity. This reprogramming/reversion of cancer cells was found in breast tumor cells, lung cancer, colon carcinoma, and melanoma [77–79]. Typically, decreasing TCTP levels led to a restructuring of the tumor architecture, where breast cancer cells formed again ductal-like structures, reminiscent of normal tissues [77]. This suggested that TCTP regulates a series of oncogenic and tumor suppressor pathways and that its silencing suppresses malignant growth [79]. These aspects of TCTP have been extensively reviewed previously [78]. Since then, it has been confirmed that TCTP is overexpressed in most tumors including clinical samples [5]. The mechanistic way, through which TCTP exerts its action, is most probably by interacting with hundreds of proteins, influencing their function, in different ways [78,80].

As TCTP is a highly conserved protein expressed in all eukaryotic organisms, some of the crucial knowledge has been generated through work in *Drosophila* by K.W. Choi and colleagues. They found that TCTP regulates the TOR pathway through interaction with the dRheb GTPase [81]. Reduction of TCTP levels led to a reduced cell number, along with a smaller organ size. Given the importance of the mTOR pathway in cancer, these results provided one of the links between TCTP and tumorigenicity. These findings in *Drosophila* were further extended by showing the implication of 14-3-3 proteins; their interaction with TCTP and Rheb is necessary for the regulation of TOR [82]. The same group demonstrated that TCTP also regulates genome stability through modulation of dATM, one of the molecular complexes implicated in the DNA damage response [83]. In addition, they showed that TCTP binds to Brahma and negatively regulates its activity [84]. Brahma is the catalytic subunit of the SW1/SNF complex, which modulates chromatin and DNA repair, and which is mutated in more than 20% of human cancers [85]. In an elegant study from Azzam's group [86], it was shown that TCTP forms complexes with ATM and γ-H2AX, suggesting a role in DNA damage and repair following exposure of human cells to low-dose γ-rays. Altogether, these data provide genetic evidence in support of the interaction of TCTP with the TOR-dependent oncogenic pathway, and of its role in maintaining genome stability.

Several studies link TCTP to epithelial to mesenchymal transition (EMT). This biological process is fundamental in early stages of embryonic development, such as the formation of the body plan during gastrulation [87]. Significant knowledge has been generated on the regulation of EMT in different developmental contexts, and evidence for its implication in cancer and metastasis is rapidly progressing [87], but still awaiting confirmation by clinical data. There has also been substantial progress in deciphering the molecular pathways involved in EMT and cancer [87]. The plasticity of cancer cells is key in EMT, especially in reprogramming somatic cells to 'stemness'. Since TCTP plays an important role in tumor reprogramming, it was speculated that it might be part of the EMT induction process. In an interesting study [88], LLC-PK1 kidney epithelial cells were used to test in vitro the potential role of TCTP for the regulation of EMT. The overexpression of TCTP enhanced migration with a reduced expression of E-cadherin and increased expression of transcription factors repressing E-cadherin, such as ZEB1, slug, or twist. Depletion of TCTP reversed the EMT phenotype and suppressed migration. Results suggested that TCTP acts through metalloproteinases, specifically MMP-9, to facilitate cell invasion [88]. In addition, the study provided data indicating that TCTP regulates pulmonary metastasis of melanoma. Another study [89], this time using A549 lung adenocarcinoma cells, suggested that TCTP was a target of TGF-β1 and necessary for EMT and cytoskeleton reorganisation. Using the same cell line, it was further confirmed that TCTP promotes EMT and tumorigenicity [90,91] by influencing the expression levels of key transcription factors (ZEB1 and alpha SMA), and of miR-200a, miR-141, and miR-424 [90]. A very recent study also showed that TCTP is a key mediator in the induction of EMT by cigarette smoke carcinogens in lung epithelial and non-small-cell lung cancer cells [70]. Overall, these recent reports on the essential role of TCTP in EMT generated substantial support for our original observation [77] that decreased cellular levels of TCTP in cancer cells inhibit tumorigenicity by interfering with migration and invasion.

The capacity of TCTP to regulate diverse pathways, from mTOR and genome integrity to cell migration, has one common feature: cell survival. This is probably best reflected in tumor biology by reshaping tumor organisation and stemness [78], the latter being a conserved TCTP function from developmental biology [92] and tissue maintenance [15] to cancer stem cells [40]. A key aspect of TCTP function as a survival factor is its interaction with both, the anti-apoptotic and pro-apoptotic machinery. It was recently discovered that TCTP contains a BH3-domain, which is a common feature in the Bcl2 family regulators of apoptosis [93]. The crystal structure of a complex of Bcl-xL with a TCTP11–31 deletion variant revealed that TCTP refolds in a helical conformation upon binding the BH3-groove of Bcl-xL. Most importantly, TCTP potentiates the anti-apoptotic function of Bcl-xL, which is a unique feature [93]. On the other hand, TCTP has been shown to interact with the p53 tumor suppressor [40,78]. TCTP and p53 are involved in a negative reciprocal feedback loop, in which p53 represses the transcription of TCTP and the latter promotes the degradation of p53 [40]. This negative feedback loop is important for cell fate.

A study in hepatocellular cancer revealed that TCTP overexpression in this cancer induces mitotic defects [41], which is in line with other data showing that TCTP is involved in stabilising the mitotic spindle and is important for orderly mitotic/meiotic progression [94]. Our data also indicate that normal breast stem cells and cancer stem cells have an increased expression of TCTP [40]. As in developmental biology, stemness has to be protected from cell death. Importantly, we observed that breast cancer patients expressing high levels of TCTP in their tumors have a high grade, aggressive malignancy with a poor prognosis [40]. Similar observations have since been published for hepatocellular cancer (HCC) [41], colorectal cancer (CRC) [95], as well as for lung [90], breast [96], and gallbladder [97] cancer. Such elevated levels of TCTP in aggressive malignant disease may contribute to yet another problem, i.e., the enhanced resistance to various treatment modalities. For breast [98], lung [99], and colorectal [100] cancer cells, it has been shown that increased TCTP levels contribute to an increase in radio- and/or chemo-resistance.

Since the beginning of our research, it was obvious that TCTP is a potentially important target for cancer treatment. Sertraline and thioridazine are able to neutralise TCTP and hence, decrease its expression, which leads to apoptosis of cancer cells [79]. Sertraline is now being tested in phase I/II clinical studies [101]. Recently, it was demonstrated that TCTP is a promising target in melanoma, also using sertraline as a drug [102]. In our initial experiments targeting TCTP, we employed anti-histaminic drugs [79]. It has since been shown that anti-histaminic drugs interact with TCTP, and they were suggested as an approach to differentiation therapy [103]. The finding that the anti-malarial drug dihydroartemisinin (DHA), which also has anti-cancer activity [104], binds to and promotes the degradation of human TCTP (fortilin) [74], prompted initial studies to test the use of artemisinin derivatives against TCTP, either alone in gallbladder cancer [97] or in combination therapy against breast cancer [96,105]. Yet another approach, i.e., TCTP-antisense oligonucleotides, is being explored as a strategy against prostate cancer, and this seems to show some promise [106].

4.2. TCTP in Cardiovascular and Metabolic Diseases

Apart from cancer, TCTP dysregulation is also involved in a range of other disease processes (reviewed in [5,107]; for a compilation see Table 3 below). In a recent study, Cai and colleagues reported that TCTP plays a pivotal role in cardiomyocyte survival [46]. Downregulation of TCTP in cardiomyocytes induced cell death with apoptotic and autophagic features. Conversely, cardiomyocyte-specific overexpression of TCTP in mice resulted in decreased susceptibility to doxorubicin-induced cardiac dysfunction. In this case, TCTP acted as a disease-preventing factor. However, in the case of atherosclerosis, two earlier publications demonstrated that TCTP promotes the disease, albeit through two quite different mechanisms. Pinkaew et al. studied the effect of heterozygous deficiency of TCTP (fortilin) in mice, in a background of hypercholesteraemia [108]. They arrived at the conclusion that TCTP prevents apoptosis and therefore the reduction of macrophages, which are main contributors to the development of atherosclerosis. Kyunglim Lee's group had previously shown that TCTP overexpression resulted in the inhibition of the Na,K-ATPase, and in the development of systemic hypertension in mice as early as six weeks after birth [109]. In a later paper, they demonstrated that, in ApoE-knockout mice, TCTP overexpression and consequently hypertension accelerated the development of atherosclerotic lesions caused by high-fat and high-cholesterol diet [110].

In a recent review article, K. Lee's group summarised the consequences of the inhibition of the Na,K-ATPase by TCTP. Apart from the development of systemic hypertension, there are additional clinically relevant consequences, i.e., an increased tendency to form lens cataracts in mice and the activation of tumorigenic signalling pathways [111]. Another cardiovascular disease, in which a direct involvement of TCTP has been documented, is pulmonary arterial hypertension (PAH), a lethal disease caused by excessive proliferation of pulmonary endothelial cells (ECs). Using a proteomic approach, Lavoie et al. identified TCTP as one of 22 proteins that are significantly altered in the blood outgrowth ECs (BOECs) of patients with hereditary PAH [112]. Immunostaining revealed a marked increase in TCTP levels particularly in complex lesions of lungs from PAH patients, as well as in a rat model of severe and irreversible PAH. TCTP-knockdown led to an increase in apoptosis and to a reduction of the hyperproliferative phenotype in BOECs from PAH patients. In a recent follow-up study, the group also observed that silencing of TCTP in such BOECs resulted in significant alterations of the morphology and the migration behaviour of these cells [113]. They also demonstrated that TCTP can be transferred from ECs to pulmonary artery smooth muscle cells via exosomes, and in this way, the protein transfers the proliferative phenotype and apoptosis resistance onto neighbouring cell layers, thus playing a core role in the pathobiology of the disease.

Since TCTP is a cytoprotective protein involved in maintaining the cellular homeostasis of specialised cell types, its dysregulation may also play a role in metabolic disease states, such as diabetes. We found that in pancreatic β-cells, TCTP levels are regulated by glucose and that TCTP participates in protecting these cells against apoptosis induced by fatty acids [114]. In a more detailed study, Tsai et al. investigated the adaptation of β-cell mass in mice, both during early development and in insulin-resistant states, and found that TCTP expression correlated with phases of β-cell proliferation and mass expansion [115]. Specific knockout of TCTP in β-cells resulted in decreased growth signalling, β-cell proliferation and mass development, eventually leading to reduced insulin production and hyperglycemia. These observations received additional support by the recent finding that in mouse pancreatic islets, TCTP expression is regulated by the insulin-response element binding protein-1 (IRE-BP1) [31].

One of the pathologies associated with diabetes is nephrotic podocyte hypertrophy, which leads to an increase of glomeruli and to proteinuria. Kim et al. used a mouse model to study the involvement of TCTP in the development of this condition [116]. They found that TCTP knockdown reduced the activation of mTORC1 downstream effectors, the overproduction of cyclin-dependent kinases, as well as the size of podocytes and the glomeruli. In *db/db* mice, knockdown of TCTP prevented the development of diabetic nephropathy. Another type of hypertrophy, not related to diabetes, where TCTP overexpression was shown to be involved, is skeletal muscle hypertrophy. An in-depth study by Goodman and colleagues elucidated several aspects of TCTP's role and regulation in skeletal muscle, using a range of mouse models [61]. They showed that TCTP is translationally up-regulated via the mTORC1 signalling pathway in skeletal muscle, under both hypertrophic and atrophic conditions.

TCTP was sufficient to induce muscle fiber hypertrophy, and the protein may also be involved in inhibiting protein degradation.

Taken together, the various examples of diseases involving TCTP dysregulation (Table 3 below) show that, depending on the specific setting or cell type, the role of TCTP can be either in preventing or in promoting disease processes. TCTP as a cytoprotective protein may be involved in preventing the development of disease, as we see in model studies for cardiomyocytes [46] or pancreatic β-cells [115]. The properties of TCTP as a growth promoting and anti-apoptotic protein can also exacerbate disease processes, e.g., by driving cells into a hyperproliferative state, as in cancer (Section 4.1), PAH [112,113], diabetic nephropathy [116], or muscle hypertrophy [61], by preventing apoptosis, as shown for atherosclerosis [108] and PAH [112,113] or by inhibiting Na,K-ATPase, which leads to hypertension [111].

4.3. *Allergic and Immune Disorders—TCTP as Histamine Releasing Factor*

The discovery of the activity of TCTP as a histamine-releasing factor (HRF) has spurred a considerable research effort aimed at delineating its specific role in triggering cellular responses associated with allergic and other immune disorders. Various aspects of this work were reviewed on earlier occasions [5,57,117,118]; however, since then considerable progress has been made in understanding certain details of the role of TCTP/HRF in the development of various allergic disorders. These recent developments are summarised in a review by T. Kawakami and colleagues in this Special Issue in *Cells* [56], and for further details, the reader is referred to this article. Here, we will just mention a few core points: 1. After some initial controversy about the IgE-binding activity of HRF, it has been clarified that HRF binds to a subset of IgE molecules, in this way triggering histamine release [117]. 2. Several lines of evidence show that it is the dimer of TCTP/HRF, which is the active form for its extracellular activity [118]. The structure of the TCTP dimer has been solved and a model for dimerisation and the IgE binding site was derived from this [119]. A more recent paper also implied the flexible loop of the TCTP/HRF dimer in the activity of the molecule in triggering cytokine release from BEAS2B cells [120]. 3. The role and involvement of HRF in the following allergic disease states has been elucidated, at least in part, either in mouse models or in limited investigations in patients (Table 3): Asthma (reviewed in [56,117,118]), atopic dermatitis [121], food allergy [122,123], and chronic urticaria [124,125]. 4. A number of peptide and other inhibitors of TCTP/HRF showed some promise in alleviating symptoms elicited by this molecule in the context of allergic diseases [56].

Table 3. Dysregulation of TCTP in disease processes.

Type of Disease	Examples	Function of TCTP	References
Cancer	*Drosophila*; HCC	TOR pathway; cell cycle progr.	[41,81,82]
	Drosophila; human cells	DNA repair/genome stability	[83,84,86]
	Breast cancer	Antagonism to p53	[40,78]
	Various cancer cells	Anti-apoptotic activity	[40,78,93]
	Breast cancer	Maintaining cancer stem cells	[40,78]
	Lung cancer, melanoma	Promoting EMT	[70,88–91]
	Melanoma, gallbladder, CRC	Involvement in metastasis	[88,95,97]
	Breast, lung, CRC cells	Radio- and chemoresistance	[98–100]
Cardiovascular and metabolic diseases	Heart failure	Protection of cardiomyocytes	[46]
	Atherosclerosis	- apoptosis in macrophages	[108]
		- causing hypertension	[110]
	Hypertension, cataracts	Inhibition of Na,K-ATPase	[109,111]
	Pulmonary arterial hypertension (PAH)	Proliferation, anti-apoptosis in epithelial/endothelial cells	[112,113]
	Hyperglycemia, diabetes	Protects pancreatic β-cells	[114,115]
	Diabetic nephropathy	Promotes podocyte growth	[116]
	Muscle hypertrophy	Increase of TCTP and mTORC1	[61]
Allergic and immune disorders	Asthma	Dimeric HRF/TCTP binds to IgE^+ on mast cells/basophils and triggers histamine and cytokine release	[56,117,118]
	Atopic dermatitis		[121]
	Food allergy		[122,123]
	Chronic urticaria		[124,125]

Another paper published in this Special Issue of *Cells* reported quite a different aspect of TCTP's involvement in inflammatory responses. The group of A. Senff-Ribeiro in Brazil had previously found that TCTP is part of the venom from the Brown Spider *Loxosceles intermedia*. They have now shown that TCTP is a synergistic factor contributing to the exacerbated inflammatory response elicited by the main toxin of the venom [126].

5. Synopsis

Recent years have seen a considerable deepening of our insight into the multiple biological functions of TCTP/fortilin/HRF, into the mechanisms of its regulation and into how dysregulation of the protein may contribute to various disease processes. At the cellular level, we learned more about the role of TCTP in the cell division process, both mitotic and meiotic; its function in stabilising polar spindle microtubules and the importance of the mitotic phosphorylation of TCTP by Plk1 for its detachment from the spindle. A novel mechanism for cell cycle control by TCTP was recently discovered in plants and insects. TCTP interacts with the CSN4 subunit of the COP9 signalosome complex to regulate cell cycle progression at the G1/S transition, through modification of the activity of Cullin–Ring ubiquitin ligases. This is important in both cell proliferation and organ development; however, the relevance of deregulation by this mechanism in diseases is yet to be documented. The importance of TCTP in organ development was further confirmed by additional examples in insects (tissue regeneration), in vertebrates (nervous system development) and in plants (lateral root formation).

The association of TCTP with the protein synthesis machinery was further consolidated by the discovery of its interaction with the additional ribosomal factors, EF1A2 and RACK1, as well as with mRNAs. Detailed structural studies confirmed that the previously reported binding of TCTP to EF1B is the most conserved interaction of TCTP. However, it is still an open question, whether TCTP affects protein synthesis rates generally, and/or modulates the translation of specific mRNAs. We certainly know several examples of TCTP regulating the stability of individual proteins.

The function of TCTP as a cytoprotective protein is well established, and several additional examples have now been reported. As a new mechanism, the involvement of TCTP in the unfolded protein response to prevent ER-stress was recently revealed. Autophagy is another important cell-homeostatic mechanism, and a few papers described the involvement of TCTP in this process as well, although the precise effect of TCTP on autophagy is still a matter of debate.

Cellular TCTP levels are highly regulated in response to alterations of a variety of cell physiologic conditions. We learned more about regulatory mechanisms that are involved in modulating TCTP levels. The list of transcription factors regulating TCTP mRNA synthesis has been extended by the tumor suppressor protein p53 and by IRE-BP1, an insulin-responsive transcription factor. We now know several translational control mechanisms, which may modulate the translational efficiency of TCTP mRNA; these are (1) signalling through the mTORC1 pathway, e.g., during growth induction of TCTP synthesis, (2) negative regulation by PKR in stress conditions, (3) mRNA stability regulation in Trypanosomes and (4) regulation through a small number of microRNAs in cancer. Several examples were reported, showing that TCTP levels may be modulated through regulated protein degradation. In serum starvation, TCTP was found to be degraded through chaperone-mediated autophagy.

The involvement of TCTP in cancer has been repeatedly proven. More examples of high TCTP levels being associated with a poor outcome in cancer patients have been reported. The participation of TCTP in the following cancer-promoting pathways has been demonstrated: the mTOR pathway and cell cycle progression, DNA repair and genome stability, antagonism to tumor suppressor p53 and anti-apoptotic activity, maintenance of 'stemness' in cancer cells, promotion of EMT and involvement in metastasis, and development of radio- and chemoresistance in cancer cells. Initial ideas to target TCTP as (part of) potential anti-cancer strategies have been published. Other disease processes, where dysregulation of TCTP might be a contributing factor, include cardiovascular diseases (arthrosclerosis and hypertension) and metabolic disorders (diabetes, muscle hypertrophy). The extracellular function of dimerised TCTP as histamine-releasing factor (HRF) in allergic and immune disorders has been further clarified, for asthma, atopic dermatitis, food allergy, and chronic urticaria.

With this extended knowledge about the principal functions of TCTP/fortilin/HRF in many biological and disease processes, our 'toolbox' should be large enough now for taking first steps towards 'translating' this knowledge into practical medical applications.

Author Contributions: U.-A.B. conceived the review; both authors wrote the article and corrected the final version. All authors have read and agreed to the published version of the manuscript.

Funding: U.-A.B. was supported by grants from the Wellcome Trust (UK) and by small grants from the Illawarra Health and Medical Research Institute and from the Graduate School of Medicine, University of Wollongong, NSW, Australia. A.T. was supported by grants from the French National Agency for Research ANR (ANR-09-BLAN-0292), the European Union Network of Excellence CONTICANET, LabEx LERMIT, INCa Projets libres de 2013-1-PLBIO-10-IGR-1.

Acknowledgments: AT thanks Robert Amson for the past present and future work done together.

Conflicts of Interest: The authors declare no conflict of interest.

References

1. Thomas, G.; Thomas, G.; Luther, H. Transcriptional and translational control of cytoplasmic proteins after serum stimulation of quiescent Swiss 3T3 cells. *Proc. Natl. Acad. Sci. USA* **1981**, *78*, 5712–5716. [CrossRef] [PubMed]
2. Yenofsky, R.; Bergmann, I.; Brawerman, G. Messenger RNA species partially in a repressed state in mouse sarcoma ascites cells. *Proc. Natl. Acad. Sci. USA* **1982**, *79*, 5876–5880. [CrossRef] [PubMed]
3. Telerman, A.E.; Amson, R.E. (Eds.) *TCTP/tpt1—Remodeling Signaling from Stem Cell to Disease*; Springer International Publishing AG: Cham, Switzerland, 2017; Volume 64, pp. 1–309.
4. Assrir, N.; Malard, F.; Lescop, E. Structural Insights into TCTP and Its Interactions with Ligands and Proteins. *Results Probl. Cell Differ.* **2017**, *64*, 9–46. [CrossRef]
5. Bommer, U.A. The Translational Controlled Tumour Protein TCTP: Biological Functions and Regulation. *Results Probl. Cell Differ.* **2017**, *64*, 69–126. [CrossRef] [PubMed]
6. Li, S.; Ge, F. Current Understanding of the TCTP Interactome. *Results Probl. Cell Differ.* **2017**, *64*, 127–136. [CrossRef]
7. Kobayashi, D.; Tokuda, T.; Sato, K.; Okanishi, H.; Nagayama, M.; Hirayama-Kurogi, M.; Ohtsuki, S.; Araki, N. Identification of a Specific Translational Machinery via TCTP-EF1A2 Interaction Regulating NF1-associated Tumor Growth by Affinity Purification and Data-independent Mass Spectrometry Acquisition (AP-DIA). *Mol. Cell Proteom.* **2019**, *18*, 245–262. [CrossRef]
8. Kubiak, J.Z.; Kloc, M. Elusive Role of TCTP Protein and mRNA in Cell Cycle and Cytoskeleton Regulation. *Results Probl. Cell Differ.* **2017**, *64*, 217–225. [CrossRef]
9. Betsch, L.; Savarin, J.; Bendahmane, M.; Szecsi, J. Roles of the Translationally Controlled Tumor Protein (TCTP) in Plant Development. *Results Probl. Cell Differ.* **2017**, *64*, 149–172. [CrossRef]
10. Choi, K.W.; Hong, S.T.; Le, T.P. Function of Translationally Controlled Tumor Protein in Organ Growth: Lessons from Drosophila Studies. *Results Probl. Cell Differ.* **2017**, *64*, 173–191. [CrossRef]
11. Roque, C.G.; Holt, C.E. Tctp in Neuronal Circuitry Assembly. *Results Probl. Cell Differ.* **2017**, *64*, 201–215. [CrossRef]
12. Jeon, H.J.; Cui, X.S.; Guo, J.; Lee, J.M.; Kim, J.S.; Oh, J.S. TCTP regulates spindle assembly during postovulatory aging and prevents deterioration in mouse oocyte quality. *Biochim. Biophys. Acta* **2017**, *1864*, 1328–1334. [CrossRef] [PubMed]
13. Betsch, L.; Boltz, V.; Brioudes, F.; Pontier, G.; Girard, V.; Savarin, J.; Wipperman, B.; Chambrier, P.; Tissot, N.; Benhamed, M.; et al. TCTP and CSN4 control cell cycle progression and development by regulating CULLIN1 neddylation in plants and animals. *PLoS Genet.* **2019**, *15*, e1007899. [CrossRef]
14. Liu, Z.L.; Xu, J.; Ling, L.; Zhang, R.; Shang, P.; Huang, Y.P. CRISPR disruption of TCTP gene impaired normal development in the silkworm Bombyx mori. *Insect Sci.* **2019**, *26*, 973–982. [CrossRef]

15. Kwon, Y.V.; Zhao, B.; Xu, C.; Lee, J.; Chen, C.L.; Vinayagam, A.; Edgar, B.A.; Perrimon, N. The role of translationally controlled tumor protein in proliferation of Drosophila intestinal stem cells. *Proc. Natl. Acad. Sci. USA* **2019**, *116*, 26591–26598. [CrossRef] [PubMed]
16. Lin, Z.; Zhang, X.; Wang, J.; Liu, W.; Liu, Q.; Ye, Y.; Dai, B.; Guo, D.; Zhang, P.; Yang, P.; et al. Translationally controlled tumor protein promotes liver regeneration by activating mTORC2/AKT signaling. *Cell Death Dis.* **2020**, *11*, 58. [CrossRef] [PubMed]
17. Roque, C.G.; Holt, C.E. Growth Cone Tctp Is Dynamically Regulated by Guidance Cues. *Front. Mol. Neurosci.* **2018**, *11*, 399. [CrossRef]
18. Roque, C.G.; Wong, H.H.; Lin, J.Q.; Holt, C.E. Tumor protein Tctp regulates axon development in the embryonic visual system. *Development* **2016**, *143*, 1134–1148. [CrossRef]
19. Bae, S.Y.; Sheverdin, V.; Maeng, J.; Lyoo, I.K.; Han, P.L.; Lee, K. Immunohistochemical Localization of Translationally Controlled Tumor Protein in Axon Terminals of Mouse Hippocampal Neurons. *Exp. Neurobiol.* **2017**, *26*, 82–89. [CrossRef]
20. Chen, S.H.; Lu, C.H.; Tsai, M.J. TCTP is Essential for Cell Proliferation and Survival during CNS Development. *Cells* **2020**, *9*, 133. [CrossRef]
21. Branco, R.; Masle, J. Systemic signalling through translationally controlled tumour protein controls lateral root formation in Arabidopsis. *J. Exp. Bot* **2019**, *70*, 3927–3940. [CrossRef]
22. Brown, M.P.; Grundy, W.N.; Lin, D.; Cristianini, N.; Sugnet, C.W.; Furey, T.S.; Ares, M., Jr.; Haussler, D. Knowledge-based analysis of microarray gene expression data by using support vector machines. *Proc. Natl. Acad. Sci. USA* **2000**, *97*, 262–267. [CrossRef] [PubMed]
23. Fleischer, T.C.; Weaver, C.M.; McAfee, K.J.; Jennings, J.L.; Link, A.J. Systematic identification and functional screens of uncharacterized proteins associated with eukaryotic ribosomal complexes. *Genes Dev.* **2006**, *20*, 1294–1307. [CrossRef] [PubMed]
24. Cans, C.; Passer, B.J.; Shalak, V.; Nancy-Portebois, V.; Crible, V.; Amzallag, N.; Allanic, D.; Tufino, R.; Argentini, M.; Moras, D.; et al. Translationally controlled tumor protein acts as a guanine nucleotide dissociation inhibitor on the translation elongation factor eEF1A. *Proc. Natl. Acad. Sci. USA* **2003**, *100*, 13892–13897. [CrossRef] [PubMed]
25. Langdon, J.M.; Vonakis, B.M.; MacDonald, S.M. Identification of the interaction between the human recombinant histamine releasing factor/translationally controlled tumor protein and elongation factor-1 delta (also known as eElongation factor-1B beta). *Biochim. Biophys. Acta* **2004**, *1688*, 232–236. [CrossRef] [PubMed]
26. Yamashita, R.; Suzuki, Y.; Takeuchi, N.; Wakaguri, H.; Ueda, T.; Sugano, S.; Nakai, K. Comprehensive detection of human terminal oligo-pyrimidine (TOP) genes and analysis of their characteristics. *Nucleic Acids Res.* **2008**, *36*, 3707–3715. [CrossRef]
27. Meyuhas, O. Synthesis of the translational apparatus is regulated at the translational level. *Eur. J. Biochem.* **2000**, *267*, 6321–6330. [CrossRef]
28. Wu, H.; Gong, W.; Yao, X.; Wang, J.; Perrett, S.; Feng, Y. Evolutionarily conserved binding of translationally controlled tumor protein to eukaryotic elongation factor 1B. *J. Biol. Chem.* **2015**, *290*, 8694–8710. [CrossRef]
29. Yao, X.; Liu, Y.J.; Cui, Q.; Feng, Y. Solution structure of a unicellular microalgae-derived translationally controlled tumor protein revealed both conserved features and structural diversity. *Arch. Biochem. Biophys.* **2019**, *665*, 23–29. [CrossRef]
30. Xie, J.; Shen, K.; Lenchine, R.V.; Gethings, L.A.; Trim, P.J.; Snel, M.F.; Zhou, Y.; Kenney, J.W.; Kamei, M.; Kochetkova, M.; et al. Eukaryotic elongation factor 2 kinase upregulates the expression of proteins implicated in cell migration and cancer cell metastasis. *Int. J. Cancer* **2018**, *142*, 1865–1877. [CrossRef]
31. Villafuerte, B.C.; Barati, M.T.; Rane, M.J.; Isaacs, S.; Li, M.; Wilkey, D.W.; Merchant, M.L. Over-expression of insulin-response element binding protein-1 (IRE-BP1) in mouse pancreatic islets increases expression of RACK1 and TCTP: Beta cell markers of high glucose sensitivity. *Biochim. Biophys. Acta* **2017**, *1865*, 186–194. [CrossRef]
32. Castello, A.; Fischer, B.; Eichelbaum, K.; Horos, R.; Beckmann, B.M.; Strein, C.; Davey, N.E.; Humphreys, D.T.; Preiss, T.; Steinmetz, L.M.; et al. Insights into RNA biology from an atlas of mammalian mRNA-binding proteins. *Cell* **2012**, *149*, 1393–1406. [CrossRef] [PubMed]
33. Nielsen, M.H.; Flygaard, R.K.; Jenner, L.B. Structural analysis of ribosomal RACK1 and its role in translational control. *Cell Signal.* **2017**, *35*, 272–281. [CrossRef]

34. Vidal, M. Role and Fate of TCTP in Protein Degradative Pathways. *Results Probl. Cell Differ.* **2017**, *64*, 137–148. [CrossRef] [PubMed]
35. Liu, H.; Peng, H.W.; Cheng, Y.S.; Yuan, H.S.; Yang-Yen, H.F. Stabilization and enhancement of the antiapoptotic activity of mcl-1 by TCTP. *Mol. Cell Biol.* **2005**, *25*, 3117–3126. [CrossRef] [PubMed]
36. Zhang, F.; Liu, B.; Wang, Z.; Yu, X.J.; Ni, Q.X.; Yang, W.T.; Mukaida, N.; Li, Y.Y. A novel regulatory mechanism of Pim-3 kinase stability and its involvement in pancreatic cancer progression. *Mol. Cancer Res.* **2013**, *11*, 1508–1520. [CrossRef] [PubMed]
37. Chattopadhyay, A.; Pinkaew, D.; Doan, H.Q.; Jacob, R.B.; Verma, S.K.; Friedman, H.; Peterson, A.C.; Kuyumcu-Martinez, M.N.; McDougal, O.M.; Fujise, K. Fortilin potentiates the peroxidase activity of Peroxiredoxin-1 and protects against alcohol-induced liver damage in mice. *Sci. Rep.* **2016**, *6*, 18701. [CrossRef]
38. Tao, J.J.; Cao, Y.R.; Chen, H.W.; Wei, W.; Li, Q.T.; Ma, B.; Zhang, W.K.; Chen, S.Y.; Zhang, J.S. Tobacco Translationally Controlled Tumor Protein Interacts with Ethylene Receptor Tobacco Histidine Kinase1 and Enhances Plant Growth through Promotion of Cell Proliferation. *Plant. Physiol.* **2015**, *169*, 96–114. [CrossRef]
39. Chen, K.; Chen, S.; Huang, C.; Cheng, H.; Zhou, R. TCTP increases stability of hypoxia-inducible factor 1alpha by interaction with and degradation of the tumour suppressor VHL. *Biol. Cell* **2013**, *105*, 208–218. [CrossRef]
40. Amson, R.; Pece, S.; Lespagnol, A.; Vyas, R.; Mazzarol, G.; Tosoni, D.; Colaluca, I.; Viale, G.; Rodrigues-Ferreira, S.; Wynendaele, J.; et al. Reciprocal repression between P53 and TCTP. *Nat. Med.* **2012**, *18*, 91–99. [CrossRef]
41. Chan, T.H.; Chen, L.; Liu, M.; Hu, L.; Zheng, B.J.; Poon, V.K.; Huang, P.; Yuan, Y.F.; Huang, J.D.; Yang, J.; et al. Translationally controlled tumor protein induces mitotic defects and chromosome missegregation in hepatocellular carcinoma development. *Hepatology* **2012**, *55*, 491–505. [CrossRef]
42. Guerrero, C.; Milenkovic, T.; Przulj, N.; Kaiser, P.; Huang, L. Characterization of the proteasome interaction network using a QTAX-based tag-team strategy and protein interaction network analysis. *Proc. Natl. Acad. Sci. USA* **2008**, *105*, 13333–13338. [CrossRef] [PubMed]
43. Rinnerthaler, M.; Lejskova, R.; Grousl, T.; Stradalova, V.; Heeren, G.; Richter, K.; Breitenbach-Koller, L.; Malinsky, J.; Hasek, J.; Breitenbach, M. Mmi1, the yeast homologue of mammalian TCTP, associates with stress granules in heat-shocked cells and modulates proteasome activity. *PLoS ONE* **2013**, *8*, e77791. [CrossRef] [PubMed]
44. Zhang, J.; Shim, G.; de Toledo, S.M.; Azzam, E.I. The Translationally Controlled Tumor Protein and the Cellular Response to Ionizing Radiation-Induced DNA Damage. *Results Probl. Cell Differ.* **2017**, *64*, 227–253. [CrossRef] [PubMed]
45. Seo, E.J.; Fischer, N.; Efferth, T. Role of TCTP for Cellular Differentiation and Cancer Therapy. *Results Probl. Cell Differ.* **2017**, *64*, 263–281. [CrossRef] [PubMed]
46. Cai, W.; Fujita, T.; Hidaka, Y.; Jin, H.; Suita, K.; Shigeta, M.; Kiyonari, H.; Umemura, M.; Yokoyama, U.; Sadoshima, J.; et al. Translationally controlled tumor protein (TCTP) plays a pivotal role in cardiomyocyte survival through a Bnip3-dependent mechanism. *Cell Death Dis.* **2019**, *10*, 549. [CrossRef] [PubMed]
47. Pinkaew, D.; Chattopadhyay, A.; King, M.D.; Chunhacha, P.; Liu, Z.; Stevenson, H.L.; Chen, Y.; Sinthujaroen, P.; McDougal, O.M.; Fujise, K. Fortilin binds IRE1alpha and prevents ER stress from signaling apoptotic cell death. *Nat. Commun.* **2017**, *8*, 18. [CrossRef]
48. De Carvalho, M.; Acencio, M.L.; Laitz, A.V.N.; de Araujo, L.M.; de Lara Campos Arcuri, M.; do Nascimento, L.C.; Maia, I.G. Impacts of the overexpression of a tomato translationally controlled tumor protein (TCTP) in tobacco revealed by phenotypic and transcriptomic analysis. *Plant. Cell Rep.* **2017**, *36*, 887–900. [CrossRef] [PubMed]
49. Jojic, B.; Amodeo, S.; Ochsenreiter, T. The translationally controlled tumor protein TCTP is involved in cell cycle progression and heat stress response in the bloodstream form of Trypanosoma brucei. *Microb. Cell* **2018**, *5*, 460–468. [CrossRef]
50. Ying, X.; Liu, Y.; Chen, L.; Bo, Q.; Xu, Q.; Li, F.; Zhou, C.; Cheng, L. Analysis of translation control tumor protein related to deltamethrin stress in Drosophila kc cells. *Chemosphere* **2019**, *231*, 450–456. [CrossRef]
51. Lee, J.S.; Jang, E.H.; Woo, H.A.; Lee, K. Regulation of Autophagy Is a Novel Tumorigenesis-Related Activity of Multifunctional Translationally Controlled Tumor Protein. *Cells* **2020**, *9*, 257. [CrossRef]

52. Chen, K.; Huang, C.; Yuan, J.; Cheng, H.; Zhou, R. Long-term artificial selection reveals a role of TCTP in autophagy in mammalian cells. *Mol. Biol. Evol.* **2014**, *31*, 2194–2211. [CrossRef] [PubMed]
53. Bae, S.Y.; Byun, S.; Bae, S.H.; Min, D.S.; Woo, H.A.; Lee, K. TPT1 (tumor protein, translationally-controlled 1) negatively regulates autophagy through the BECN1 interactome and an MTORC1-mediated pathway. *Autophagy* **2017**, *13*, 820–833. [CrossRef] [PubMed]
54. Vojtova, J.; Hasek, J. Mmi1, the Yeast Ortholog of Mammalian Translationally Controlled Tumor Protein (TCTP), Negatively Affects Rapamycin-Induced Autophagy in Post-Diauxic Growth Phase. *Cells* **2020**, *9*, 138. [CrossRef] [PubMed]
55. MacDonald, S.M.; Rafnar, T.; Langdon, J.; Lichtenstein, L.M. Molecular identification of an IgE-dependent histamine-releasing factor. *Science* **1995**, *269*, 688–690. [CrossRef] [PubMed]
56. Kawakami, Y.; Kasakura, K.; Kawakami, T. Histamine-Releasing Factor, a New Therapeutic Target in Allergic Diseases. *Cells* **2019**, *8*, 1515. [CrossRef]
57. MacDonald, S.M. History of Histamine-Releasing Factor (HRF)/Translationally Controlled Tumor Protein (TCTP) Including a Potential Therapeutic Target in Asthma and Allergy. *Results Probl. Cell Differ.* **2017**, *64*, 291–308. [CrossRef]
58. Andree, H.; Thiele, H.; Fahling, M.; Schmidt, I.; Thiele, B.J. Expression of the human TPT1 gene coding for translationally controlled tumor protein (TCTP) is regulated by CREB transcription factors. *Gene* **2006**, *380*, 95–103. [CrossRef]
59. Schmidt, I.; Fahling, M.; Nafz, B.; Skalweit, A.; Thiele, B.J. Induction of translationally controlled tumor protein (TCTP) by transcriptional and post-transcriptional mechanisms. *FEBS J.* **2007**, *274*, 5416–5424. [CrossRef]
60. Bommer, U.A.; Iadevaia, V.; Chen, J.; Knoch, B.; Engel, M.; Proud, C.G. Growth-factor dependent expression of the translationally controlled tumour protein TCTP is regulated through the PI3-K/Akt/mTORC1 signalling pathway. *Cell Signal.* **2015**, *27*, 1557–1568. [CrossRef]
61. Goodman, C.A.; Coenen, A.M.; Frey, J.W.; You, J.S.; Barker, R.G.; Frankish, B.P.; Murphy, R.M.; Hornberger, T.A. Insights into the role and regulation of TCTP in skeletal muscle. *Oncotarget* **2017**, *8*, 18754–18772. [CrossRef]
62. Bommer, U.A.; Borovjagin, A.V.; Greagg, M.A.; Jeffrey, I.W.; Russell, P.; Laing, K.G.; Lee, M.; Clemens, M.J. The mRNA of the translationally controlled tumor protein P23/TCTP is a highly structured RNA, which activates the dsRNA-dependent protein kinase PKR. *RNA* **2002**, *8*, 478–496. [CrossRef]
63. Bommer, U.A.; Heng, C.; Perrin, A.; Dash, P.; Lobov, S.; Elia, A.; Clemens, M.J. Roles of the translationally controlled tumour protein (TCTP) and the double-stranded RNA-dependent protein kinase, PKR, in cellular stress responses. *Oncogene* **2010**, *29*, 763–773. [CrossRef]
64. Lackey, L.; Coria, A.; Woods, C.; McArthur, E.; Laederach, A. Allele-specific SHAPE-MaP assessment of the effects of somatic variation and protein binding on mRNA structure. *RNA* **2018**, *24*, 513–528. [CrossRef]
65. Thiele, H.; Berger, M.; Skalweit, A.; Thiele, B.J. Expression of the gene and processed pseudogenes encoding the human and rabbit translationally controlled tumour protein (TCTP). *Eur. J. Biochem.* **2000**, *267*, 5473–5481. [CrossRef]
66. Jojic, B.; Amodeo, S.; Bregy, I.; Ochsenreiter, T. Distinct 3′ UTRs regulate the life-cycle-specific expression of two TCTP paralogs in Trypanosoma brucei. *J. Cell Sci.* **2018**, *131*. [CrossRef]
67. Gaken, J.; Mohamedali, A.M.; Jiang, J.; Malik, F.; Stangl, D.; Smith, A.E.; Chronis, C.; Kulasekararaj, A.G.; Thomas, N.S.; Farzaneh, F.; et al. A functional assay for microRNA target identification and validation. *Nucleic Acids Res.* **2012**, *40*, e75. [CrossRef]
68. Lo, W.Y.; Wang, H.J.; Chiu, C.W.; Chen, S.F. miR-27b-regulated TCTP as a novel plasma biomarker for oral cancer: From quantitative proteomics to post-transcriptional study. *J. Proteom.* **2012**, *77*, 154–166. [CrossRef] [PubMed]
69. Jian, M.; Du, Q.; Zhu, D.; Mao, Z.; Wang, X.; Feng, Y.; Xiao, Z.; Wang, H.; Zhu, Y. Tumor suppressor miR-145-5p sensitizes prolactinoma to bromocriptine by downregulating TPT1. *J. Endocrinol. Investig.* **2018**. [CrossRef] [PubMed]
70. Liu, L.Z.; Wang, M.; Xin, Q.; Wang, B.; Chen, G.G.; Li, M.Y. The permissive role of TCTP in PM2.5/NNK-induced epithelial-mesenchymal transition in lung cells. *J. Transl. Med.* **2020**, *18*, 66. [CrossRef] [PubMed]

71. Zhang, D.; Li, F.; Weidner, D.; Mnjoyan, Z.H.; Fujise, K. Physical and functional interaction between myeloid cell leukemia 1 protein (MCL1) and Fortilin. The potential role of MCL1 as a fortilin chaperone. *J. Biol. Chem.* **2002**, *277*, 37430–37438. [CrossRef]
72. Baylot, V.; Katsogiannou, M.; Andrieu, C.; Taieb, D.; Acunzo, J.; Giusiano, S.; Fazli, L.; Gleave, M.; Garrido, C.; Rocchi, P. Targeting TCTP as a new therapeutic strategy in castration-resistant prostate cancer. *Mol. Ther.* **2012**, *20*, 2244–2256. [CrossRef] [PubMed]
73. Baylot, V.; Karaki, S.; Rocchi, P. TCTP Has a Crucial Role in the Different Stages of Prostate Cancer Malignant Progression. *Results Probl. Cell Differ.* **2017**, *64*, 255–261. [CrossRef] [PubMed]
74. Fujita, T.; Felix, K.; Pinkaew, D.; Hutadilok-Towatana, N.; Liu, Z.; Fujise, K. Human fortilin is a molecular target of dihydroartemisinin. *FEBS Lett.* **2008**, *582*, 1055–1060. [CrossRef] [PubMed]
75. Kubiak, J.Z.; Bazile, F.; Pascal, A.; Richard-Parpaillon, L.; Polanski, Z.; Ciemerych, M.A.; Chesnel, F. Temporal regulation of embryonic M-phases. *Folia Histochem. Cytobiol.* **2008**, *46*, 5–9. [CrossRef] [PubMed]
76. Bonhoure, A.; Vallentin, A.; Martin, M.; Senff-Ribeiro, A.; Amson, R.; Telerman, A.; Vidal, M. Acetylation of translationally controlled tumor protein promotes its degradation through chaperone-mediated autophagy. *Eur. J. Cell Biol.* **2017**, *96*, 83–98. [CrossRef] [PubMed]
77. Tuynder, M.; Susini, L.; Prieur, S.; Besse, S.; Fiucci, G.; Amson, R.; Telerman, A. Biological models and genes of tumor reversion: Cellular reprogramming through tpt1/TCTP and SIAH-1. *Proc. Natl. Acad. Sci. USA* **2002**, *99*, 14976–14981. [CrossRef] [PubMed]
78. Amson, R.; Pece, S.; Marine, J.C.; Di Fiore, P.P.; Telerman, A. TPT1/ TCTP-regulated pathways in phenotypic reprogramming. *Trends Cell Biol.* **2013**, *23*, 37–46. [CrossRef]
79. Tuynder, M.; Fiucci, G.; Prieur, S.; Lespagnol, A.; Geant, A.; Beaucourt, S.; Duflaut, D.; Besse, S.; Susini, L.; Cavarelli, J.; et al. Translationally controlled tumor protein is a target of tumor reversion. *Proc. Natl. Acad. Sci. USA* **2004**, *101*, 15364–15369. [CrossRef] [PubMed]
80. Li, S.; Chen, M.; Xiong, Q.; Zhang, J.; Cui, Z.; Ge, F. Characterization of the Translationally Controlled Tumor Protein (TCTP) Interactome Reveals Novel Binding Partners in Human Cancer Cells. *J. Proteome Res.* **2016**, *15*, 3741–3751. [CrossRef]
81. Hsu, Y.C.; Chern, J.J.; Cai, Y.; Liu, M.; Choi, K.W. Drosophila TCTP is essential for growth and proliferation through regulation of dRheb GTPase. *Nature* **2007**, *445*, 785–788. [CrossRef]
82. Le, T.P.; Vuong, L.T.; Kim, A.R.; Hsu, Y.C.; Choi, K.W. 14-3-3 proteins regulate Tctp-Rheb interaction for organ growth in Drosophila. *Nat. Commun.* **2016**, *7*, 11501. [CrossRef] [PubMed]
83. Hong, S.T.; Choi, K.W. TCTP directly regulates ATM activity to control genome stability and organ development in Drosophila melanogaster. *Nat. Commun.* **2013**, *4*, 2986. [CrossRef]
84. Hong, S.T.; Choi, K.W. Antagonistic roles of Drosophila Tctp and Brahma in chromatin remodelling and stabilizing repeated sequences. *Nat. Commun.* **2016**, *7*, 12988. [CrossRef]
85. Kadoch, C.; Crabtree, G.R. Mammalian SWI/SNF chromatin remodeling complexes and cancer: Mechanistic insights gained from human genomics. *Sci. Adv.* **2015**, *1*, e1500447. [CrossRef] [PubMed]
86. Zhang, J.; de Toledo, S.M.; Pandey, B.N.; Guo, G.; Pain, D.; Li, H.; Azzam, E.I. Role of the translationally controlled tumor protein in DNA damage sensing and repair. *Proc. Natl. Acad. Sci. USA* **2012**, *109*, E926–E933. [CrossRef] [PubMed]
87. Nieto, M.A.; Huang, R.Y.; Jackson, R.A.; Thiery, J.P. Emt: 2016. *Cell* **2016**, *166*, 21–45. [CrossRef]
88. Bae, S.Y.; Kim, H.J.; Lee, K.J.; Lee, K. Translationally controlled tumor protein induces epithelial to mesenchymal transition and promotes cell migration, invasion and metastasis. *Sci. Rep.* **2015**, *5*, 8061. [CrossRef]
89. Mishra, D.K.; Srivastava, P.; Sharma, A.; Prasad, R.; Bhuyan, S.K.; Malage, R.; Kumar, P.; Yadava, P.K. Translationally controlled tumor protein (TCTP) is required for TGF-beta1 induced epithelial to mesenchymal transition and influences cytoskeletal reorganization. *Biochim. Biophys. Acta Mol. Cell Res.* **2018**, *1865*, 67–75. [CrossRef]
90. Sun, R.; Lu, X.; Gong, L.; Jin, F. TCTP promotes epithelial-mesenchymal transition in lung adenocarcinoma. *Onco Targets Ther.* **2019**, *12*, 1641–1653. [CrossRef]
91. Wang, L.; Tang, Y.; Zhao, M.; Mao, S.; Wu, L.; Liu, S.; Liu, D.; Zhao, G.; Wang, X. Knockdown of translationally controlled tumor protein inhibits growth, migration and invasion of lung cancer cells. *Life Sci.* **2018**, *193*, 292–299. [CrossRef]

92. Koziol, M.J.; Garrett, N.; Gurdon, J.B. Tpt1 activates transcription of oct4 and nanog in transplanted somatic nuclei. *Curr. Biol.* **2007**, *17*, 801–807. [CrossRef] [PubMed]
93. Thebault, S.; Agez, M.; Chi, X.; Stojko, J.; Cura, V.; Telerman, S.B.; Maillet, L.; Gautier, F.; Billas-Massobrio, I.; Birck, C.; et al. TCTP contains a BH3-like domain, which instead of inhibiting, activates Bcl-xL. *Sci. Rep.* **2016**, *6*, 19725. [CrossRef] [PubMed]
94. Jeon, H.J.; You, S.Y.; Park, Y.S.; Chang, J.W.; Kim, J.S.; Oh, J.S. TCTP regulates spindle microtubule dynamics by stabilizing polar microtubules during mouse oocyte meiosis. *Biochim. Biophys. Acta* **2016**, *1863*, 630–637. [CrossRef]
95. Xiao, B.; Chen, D.; Luo, S.; Hao, W.; Jing, F.; Liu, T.; Wang, S.; Geng, Y.; Li, L.; Xu, W.; et al. Extracellular translationally controlled tumor protein promotes colorectal cancer invasion and metastasis through Cdc42/JNK/ MMP9 signaling. *Oncotarget* **2016**, *7*, 50057–50073. [CrossRef]
96. Lucibello, M.; Adanti, S.; Antelmi, E.; Dezi, D.; Ciafre, S.; Carcangiu, M.L.; Zonfrillo, M.; Nicotera, G.; Sica, L.; De Braud, F.; et al. Phospho-TCTP as a therapeutic target of Dihydroartemisinin for aggressive breast cancer cells. *Oncotarget* **2015**, *6*, 5275–5291. [CrossRef] [PubMed]
97. Zhang, F.; Ma, Q.; Xu, Z.; Liang, H.; Li, H.; Ye, Y.; Xiang, S.; Zhang, Y.; Jiang, L.; Hu, Y.; et al. Dihydroartemisinin inhibits TCTP-dependent metastasis in gallbladder cancer. *J. Exp. Clin. Cancer Res.* **2017**, *36*, 68. [CrossRef] [PubMed]
98. Jung, J.; Lee, J.S.; Lee, Y.S.; Lee, K. Radiosensitivity of Cancer Cells Is Regulated by Translationally Controlled Tumor Protein. *Cancers* **2019**, *11*, 386. [CrossRef]
99. Du, J.; Yang, P.; Kong, F.; Liu, H. Aberrant expression of translationally controlled tumor protein (TCTP) can lead to radioactive susceptibility and chemosensitivity in lung cancer cells. *Oncotarget* **2017**, *8*, 101922–101935. [CrossRef]
100. Bommer, U.A.; Vine, K.L.; Puri, P.; Engel, M.; Belfiore, L.; Fildes, K.; Batterham, M.; Lochhead, A.; Aghmesheh, M. Translationally controlled tumour protein TCTP is induced early in human colorectal tumours and contributes to the resistance of HCT116 colon cancer cells to 5-FU and oxaliplatin. *Cell Commun. Signal.* **2017**, *15*, 9. [CrossRef]
101. Amson, R.; Auclair, C.; Andre, F.; Karp, J.; Telerman, A. Targeting TCTP with Sertraline and Thioridazine in Cancer Treatment. *Results Probl. Cell Differ.* **2017**, *64*, 283–290. [CrossRef]
102. Boia-Ferreira, M.; Basilio, A.B.; Hamasaki, A.E.; Matsubara, F.H.; Appel, M.H.; Da Costa, C.R.V.; Amson, R.; Telerman, A.; Chaim, O.M.; Veiga, S.S.; et al. TCTP as a therapeutic target in melanoma treatment. *Br. J. Cancer* **2017**, *117*, 656–665. [CrossRef] [PubMed]
103. Seo, E.J.; Efferth, T. Interaction of antihistaminic drugs with human translationally controlled tumor protein (TCTP) as novel approach for differentiation therapy. *Oncotarget* **2016**, *7*, 16818–16839. [CrossRef]
104. Efferth, T. Molecular pharmacology and pharmacogenomics of artemisinin and its derivatives in cancer cells. *Curr. Drug Targets* **2006**, *7*, 407–421. [CrossRef] [PubMed]
105. D'Amico, S.; Krasnowska, E.K.; Manni, I.; Toietta, G.; Baldari, S.; Piaggio, G.; Ranalli, M.; Gambacurta, A.; Vernieri, C.; Di Giacinto, F.; et al. DHA Affects Microtubule Dynamics Through Reduction of Phospho-TCTP Levels and Enhances the Antiproliferative Effect of T-DM1 in Trastuzumab-Resistant HER2-Positive Breast Cancer Cell Lines. *Cells* **2020**, *9*, 1260. [CrossRef]
106. Karaki, S.; Benizri, S.; Mejias, R.; Baylot, V.; Branger, N.; Nguyen, T.; Vialet, B.; Oumzil, K.; Barthelemy, P.; Rocchi, P. Lipid-oligonucleotide conjugates improve cellular uptake and efficiency of TCTP-antisense in castration-resistant prostate cancer. *J. Control. Release* **2017**, *258*, 1–9. [CrossRef] [PubMed]
107. Pinkaew, D.; Fujise, K. Fortilin: A Potential Target for the Prevention and Treatment of Human Diseases. *Adv. Clin. Chem.* **2017**, *82*, 265–300. [CrossRef]
108. Pinkaew, D.; Le, R.J.; Chen, Y.; Eltorky, M.; Teng, B.B.; Fujise, K. Fortilin reduces apoptosis in macrophages and promotes atherosclerosis. *Am. J. Physiol. Heart Circ. Physiol.* **2013**, *305*, H1519–H1529. [CrossRef]
109. Kim, M.J.; Kwon, J.S.; Suh, S.H.; Suh, J.K.; Jung, J.; Lee, S.N.; Kim, Y.H.; Cho, M.C.; Oh, G.T.; Lee, K. Transgenic overexpression of translationally controlled tumor protein induces systemic hypertension via repression of Na+,K+-ATPase. *J. Mol. Cell Cardiol.* **2008**, *44*, 151–159. [CrossRef]
110. Cho, Y.; Maeng, J.; Ryu, J.; Shin, H.; Kim, M.; Oh, G.T.; Lee, M.Y.; Lee, K. Hypertension resulting from overexpression of translationally controlled tumor protein increases the severity of atherosclerosis in apolipoprotein E knock-out mice. *Transgenic Res.* **2012**, *21*, 1245–1254. [CrossRef]

111. Jung, J.; Ryu, S.; Ki, I.A.; Woo, H.A.; Lee, K. Some Biological Consequences of the Inhibition of Na,K-ATPase by Translationally Controlled Tumor Protein (TCTP). *Int. J. Mol. Sci.* **2018**, *19*, 1657. [CrossRef]
112. Lavoie, J.R.; Ormiston, M.L.; Perez-Iratxeta, C.; Courtman, D.W.; Jiang, B.; Ferrer, E.; Caruso, P.; Southwood, M.; Foster, W.S.; Morrell, N.W.; et al. Proteomic analysis implicates translationally controlled tumor protein as a novel mediator of occlusive vascular remodeling in pulmonary arterial hypertension. *Circulation* **2014**, *129*, 2125–2135. [CrossRef] [PubMed]
113. Ferrer, E.; Dunmore, B.J.; Hassan, D.; Ormiston, M.L.; Moore, S.; Deighton, J.; Long, L.; Yang, X.D.; Stewart, D.J.; Morrell, N.W. A Potential Role for Exosomal Translationally Controlled Tumor Protein Export in Vascular Remodeling in Pulmonary Arterial Hypertension. *Am. J. Respir. Cell Mol. Biol.* **2018**, *59*, 467–478. [CrossRef] [PubMed]
114. Diraison, F.; Hayward, K.; Sanders, K.L.; Brozzi, F.; Lajus, S.; Hancock, J.; Francis, J.E.; Ainscow, E.; Bommer, U.A.; Molnar, E.; et al. Translationally controlled tumour protein (TCTP) is a novel glucose-regulated protein that is important for survival of pancreatic beta cells. *Diabetologia* **2011**, *54*, 368–379. [CrossRef] [PubMed]
115. Tsai, M.J.; Yang-Yen, H.F.; Chiang, M.K.; Wang, M.J.; Wu, S.S.; Chen, S.H. TCTP is essential for beta-cell proliferation and mass expansion during development and beta-cell adaptation in response to insulin resistance. *Endocrinology* **2014**, *155*, 392–404. [CrossRef] [PubMed]
116. Kim, D.K.; Nam, B.Y.; Li, J.J.; Park, J.T.; Lee, S.H.; Kim, D.H.; Kim, J.Y.; Kang, H.Y.; Han, S.H.; Yoo, T.H.; et al. Translationally controlled tumour protein is associated with podocyte hypertrophy in a mouse model of type 1 diabetes. *Diabetologia* **2012**, *55*, 1205–1217. [CrossRef]
117. Kawakami, T.; Kashiwakura, J.; Kawakami, Y. Histamine-releasing factor and immunoglobulins in asthma and allergy. *Allergy Asthma Immunol. Res.* **2014**, *6*, 6–12. [CrossRef]
118. Kim, M.; Maeng, J.; Lee, K. Dimerization of TCTP and its clinical implications for allergy. *Biochimie* **2013**, *95*, 659–666. [CrossRef]
119. Dore, K.A.; Kashiwakura, J.I.; McDonnell, J.M.; Gould, H.J.; Kawakami, T.; Sutton, B.J.; Davies, A.M. Crystal structures of murine and human Histamine-Releasing Factor (HRF/TCTP) and a model for HRF dimerisation in mast cell activation. *Mol. Immunol.* **2018**, *93*, 216–222. [CrossRef]
120. Lee, H.; Kim, M.S.; Lee, J.S.; Cho, H.; Park, J.; Hae Shin, D.; Lee, K. Flexible loop and helix 2 domains of TCTP are the functional domains of dimerized TCTP. *Sci. Rep.* **2020**, *10*, 197. [CrossRef]
121. Jin, X.H.; Lim, J.; Shin, D.H.; Maeng, J.; Lee, K. Dimerized Translationally Controlled Tumor Protein-Binding Peptide Ameliorates Atopic Dermatitis in NC/Nga Mice. *Int. J. Mol. Sci.* **2017**, *18*, 256. [CrossRef]
122. Ando, T.; Kashiwakura, J.I.; Itoh-Nagato, N.; Yamashita, H.; Baba, M.; Kawakami, Y.; Tsai, S.H.; Inagaki, N.; Takeda, K.; Iwata, T.; et al. Histamine-releasing factor enhances food allergy. *J. Clin. Investig.* **2017**, *127*, 4541–4553. [CrossRef] [PubMed]
123. Kashiwakura, J.I.; Ando, T.; Karasuyama, H.; Kubo, M.; Matsumoto, K.; Matsuda, T.; Kawakami, T. The basophil-IL-4-mast cell axis is required for food allergy. *Allergy* **2019**, *74*, 1992–1996. [CrossRef] [PubMed]
124. Huang, X.; Li, Z.; Sun, R. Synergistic Actions of Histamine-Releasing Factor and Histamine Releasing Factor-Reactive IgE in Chronic Urticaria. *Int. Arch. Allergy Immunol.* **2017**, *172*, 27–32. [CrossRef] [PubMed]
125. Ulambayar, B.; Lee, H.; Yang, E.M.; Park, H.S.; Lee, K.; Ye, Y.M. Dimerized, Not Monomeric, Translationally Controlled Tumor Protein Induces Basophil Activation and Mast Cell Degranulation in Chronic Urticaria. *Immune Netw.* **2019**, *19*, e20. [CrossRef]
126. Boia-Ferreira, M.; Moreno, K.G.; Basilio, A.B.C.; da Silva, L.P.; Vuitika, L.; Soley, B.; Wille, A.C.M.; Donatti, L.; Barbaro, K.C.; Chaim, O.M.; et al. TCTP from Loxosceles Intermedia (Brown Spider) Venom Contributes to the Allergic and Inflammatory Response of Cutaneous Loxoscelism. *Cells* **2019**, *8*, 1489. [CrossRef]

cells

MDPI

Review

Histamine-Releasing Factor, a New Therapeutic Target in Allergic Diseases

Yu Kawakami [1], Kazumi Kasakura [1] and Toshiaki Kawakami [1,2,*]

1 Division of Cell Biology, La Jolla Institute for Immunology; La Jolla, CA 92037, USA; ykawakami@lji.org (Y.K.); kkasakura@lji.org (K.K.)
2 Department of Dermatology, School of Medicine, University of California San Diego, La Jolla, CA 92037, USA
* Correspondence: toshi@lji.org; Tel.: +85-8-752-6814

Received: 30 October 2019; Accepted: 21 November 2019; Published: 26 November 2019

Abstract: Histamine-releasing activities on human basophils have been studied as potential allergy-causing agents for four decades. An IgE-dependent histamine-releasing factor (HRF) was recently shown to interact with a subset of immunoglobulins. Peptides or recombinant proteins that block the interactions between HRF and IgE have emerged as promising anti-allergic therapeutics, as administration of them prevented or ameliorated type 2 inflammation in animal models of allergic diseases such as asthma and food allergy. Basic and clinical studies support the notion that HRF amplifies IgE-mediated activation of mast cells and basophils. We discuss how secreted HRF promotes allergic inflammation in vitro and in vivo complex disease settings.

Keywords: allergy; mast cells; basophils; IgE; FcεRI; HRF; translationally controlled tumor protein (TCTP)

1. Introduction

Activation of mast cells and basophils via high-affinity IgE receptors (FcεRI) on the cell surface plays an essential role in allergic reactions. Multivalent allergens induce the aggregation or cross-linking of IgE-bound FcεRI to trigger their activation [1]. Activated mast cells and basophils release preformed chemicals (e.g., histamine, serotonin) and protein inflammatory mediators (e.g., proteases, tumor necrosis factor (TNF)), and de novo synthesize and secrete arachidonic acid-derived lipids, cytokines, chemokines, and growth factors [2,3]. These factors promote type 2 inflammation in allergic individuals. In this review, we will discuss histamine-releasing factor (HRF)-mediated regulation of mast cell/basophil activation via FcεRI and its roles in allergic and other immune diseases.

2. What Is HRF?

Cytokine-like factors able to activate basophils in body fluids of allergic patients have been studied for many years [4]. Several chemokines were shown to induce histamine release from human basophils in an IgE-independent manner [5–7]. On the other hand, an IgE-dependent factor with histamine-releasing activity (HRF) was molecularly cloned by Susan MacDonald's group in 1995 [8]. Coincidentally, HRF happened to be identical to the protein termed translationally-controlled tumor protein (TCTP), fortilin, p21, and p23. It is often referred to as TCTP intracellularly and is required for cell cycle progression, proliferation, survival, and malignant transformation in a variety of cell types [9–14]. Extracellularly referred to as HRF (we follow this convention in this manuscript), it is an evolutionally conserved protein (96% identical between human and mouse proteins) composed of 172 amino acids with no known related proteins. Human HRF/TCTP is encoded by the *TPT1* gene on chromosome 13. Although numerous single nucleotide polymorphisms (SNPs) are associated with allergic diseases, no genetic associations with gene expression (eQTLs) are found in the *TPT1* locus (http://dicew-database.org). Similar to antigen/IgE-mediated activation, HRF induces not only histamine release, but also IL-4 and IL-13 secretion from human basophils and IL-13 and TNF secretion from murine mast cells [15,16]. Despite the lack of a signal sequence, it is secreted as a cargo of extracellular vesicles (EVs), particularly in exosomes [17–20]. Intriguingly, the responsiveness of basophils to HRF depends on a particular type of IgE; IgE derived from certain atopic patients, termed IgE^+, can prime basophils in response to HRF, but other IgE molecules, termed IgE^-, are unable to do so [21]. The dichotomy of IgE^+ vs. IgE^- was discovered long before the molecular cloning of HRF, and several possibilities exist to explain the heterogeneity of IgE molecules: 1) structural differences in the constant regions of IgE, for example, by differences in glycosylation or alternative mRNA splicing at the ε chain 3′ terminal region [22]; 2) IgE^+ being an HRF-specific IgE antibody, that is, HRF acting as an IgE autoantigen; 3) IgE^+ reactivity due to the presence of anti-IgE antibodies in the serum.

In contrast to an earlier report suggesting that HRF does not bind to IgE [23], Kashiwakura et al. showed that a subset of IgE and IgG molecules are able to directly bind to HRF via two Ig Fab-interacting sites: the N-terminal 19 residue stretch (N19) and the H3 helix [24]. These observations are in line with an earlier speculation that the dichotomy of IgE^+ vs. IgE^- may be caused by differences in IgE variable region subgroups [25]. However, another speculation that IgE^+ reactivity is related to glycosylation of IgE [21] was not supported by the observation that mannose-specific lectins could not distinguish between basophils sensitized with IgE^+ or with IgE^- [26]. Despite these studies, it still remains possible that glycosylation at V_H and V_L regions might contribute to the IgE^+ reactivity. In light of recent revelations regarding IgE glycosylation [27], the potential role of glycosylation may be worth revisiting.

3. Bioactive Forms of HRF

HRF is constitutively secreted as a monomer, a disulfide-linked dimer, and higher molecular weight oligomers. Crystal structures of HRF monomers from various species and a homodimer of human HRF have been solved. The homodimer is made by a disulfide bond through a Cys172-Cys172 linkage between two monomers [28,29]. Kim et al. showed that N-terminally truncated recombinant rat HRF proteins, Del-N11TCTP and Del-N35TCTP, but not full-length TCTP, also form disulfide-linked dimers with strong cytokine-like activity [29]. However, Doré et al. observed dimers of full-length mouse and human HRFs [28]. Consistent with the efficacy of HRF inhibitors in allergic disease models (see below), IgE-binding sequences (i.e., N19 and H3) are exposed on the molecular surface of HRF dimer (Figure 1a,b) [28]. Recombinant HRF homodimers, but not monomers, synthesized in *E. coli* can activate murine mast cells [30]. GST-HRF fusion proteins induce not only histamine release [8] but also secretion of IL-4 and IL-13 from human basophils [15,16]. It is well known that GST fusion proteins can form dimers. Thus, these results suggest that FcεRI-bound IgE molecules are cross-linked by HRF dimers (Figure 1c). HRF homodimers are also able to enhance IgE and antigen-stimulated production of IL-6, IL-13, and TNF but not β-hexosaminidase release (which is fully activated by stimulation with antigen) from murine mast cells. This result suggests that cytokine production requires stronger and/or more persistent FcεRI cross-linking than does degranulation. These observations can be extended to the argument that HRF exerts its effects by activating FcεRI signaling pathways. However, subtle differences in signaling may occur, as components of the ligand complex are different when cells are stimulated with antigen/IgE complexes bound to FcεRI with or without HRF. Intranasal instillation of recombinant HRF (including HRF dimers), but not HRF-2CA (a monomeric mutant of HRF with the two cysteine residues being replaced with alanine), reduced/carboxymethylated or boiled HRF, in naïve mice triggered airway inflammation in an FcεRI-dependent manner [24]. The wide gamut of signs seen in allergic diseases ranging from the mild skin rashes and gastrointestinal symptoms, to more severe signs such as pulmonary distress and systemic anaphylaxis, could be due to different levels of contributions of HRF dimer/oligomers as well as other factors such as variable antigen valencies and concentrations or FcεRI occupancy by antigen-specific IgE. Further analysis of HRF regulation of

FcεRI activation is warranted to understand how different forms of HRF affect allergen/IgE-mediated FcεRI cross-linking.

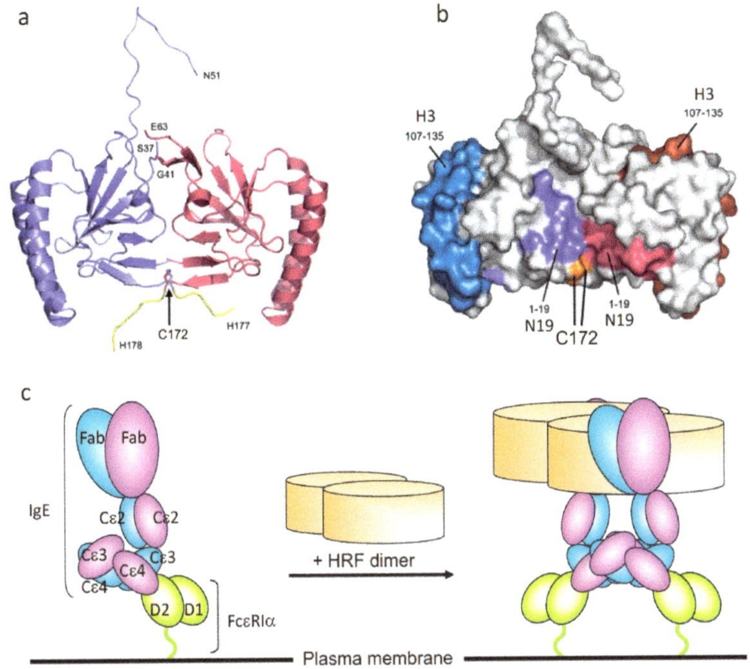

Figure 1. The crystal structure of histamine-releasing factor (HRF) dimer and HRF dimer/IgE-mediated FcεRI crosslinking. (**a**) Overall structure of a human HRF dimer. The two molecules of the asymmetric unit are colored blue and pink. The C-terminal tag is colored yellow, and the positions of C-terminal residues and residues adjacent to the disordered loop are indicated. (**b**) The two monomers of the HRF dimer are colored white and Cys172 is colored orange. For the first monomer, the two IgE binding sites, mapped to residues Met1–Lys19 (N19), and Arg107–Ile135 (H3), are colored light blue and dark blue, respectively. For the second monomer, residues 1-19 (N19) and 107-135 (H3) are colored light and dark pink, respectively. (**c**) Model for HRF dimer/IgE-mediated FcεRI crosslinking. IgE binds FcεRI α chain via the interaction between IgE–Cε3 and FcεRIα–D2 domains. One HRF molecule can bind one (this version depicted) or two molecules of IgE via interactions with the N19 and H3 regions of HRF. After binding of an HRF dimer, two (this version depicted) or four FcεRI α chain-nucleated complexes will be formed (Right). The cytoplasmic portion of FcεRI α as well as β and γ chains of FcεRI are omitted for clarity.

4. HRF in Allergic and Immune Diseases

Allergic diseases such as atopic dermatitis, food allergy, asthma, and allergic rhinitis are type 2 inflammatory diseases in allergen-sensitized individuals with organ-specific or systemic disease susceptibility [31–33]. Type 2 inflammation is caused by type 2 innate lymphoid cells, allergen-specific Th2 cells, and epithelial-derived cytokine- and Th2 cytokine-recruited mast cells and eosinophils [34–37]. HRF secretion was found in nasal, skin blister, and bronchoalveolar lavage fluids during the late phase of allergic reactions [38], implicating HRF in allergic diseases (Table 1). Long before the molecular nature of HRF was revealed, a study showed that patients with food allergy and atopic dermatitis, but not patients with atopic dermatitis alone, have higher rates of spontaneous release of histamine from basophils than normal subjects [39], implying HRF's involvement in food allergy. However, definitive evidence for pathological roles of HRF in allergy had been elusive until recently, as there were intractable obstacles in HRF research: (i) HRF/TCTP has both intracellular and extracellular functions, but no tools were available to dissect these functions in complex in vivo settings. (ii) Despite considerable efforts, researchers were unable to identify an HRF receptor for many years [23]. (iii) HRF knockout mice were embryonically lethal [40–42], thus severely limiting in vivo functional studies. As described above, Kashiwakura et al. identified a subset of IgE and IgG molecules as HRF receptors [24]: mapping of the Ig Fab-binding

Table 1. HRF in allergic and immune disorders.

Disease	Modulation of Animal Disease Models by HRF or HRF Inhibitors	Human Patients
Asthma	↓OVA-induced airway inflammation by HRF inhibitors (N19, H3) ↓*Aspergillus fumigatus*-induced airway inflammation by HRF inhibitors (N19) ↑airway inflammation induced by intranasal instillation of recombinant HRF ↓OVA-induced airway inflammation by dTBP2 peptide	
Atopic dermatitis (AD)	↓passive cutaneous anaphylaxis by HRF inhibitors (N19) ↓house dust mite allergen-induced skin inflammation in NC/Nga mice by dTBP2 peptide	↑serum HRF, ↑serum HRF-reactive IgE
Food allergy (FA)	**OVA-induced FA:** ↑serum HRF-reactive IgE, ↑HRF dimer/ oligomers in jejunum, ↑diarrhea, ↑hypothermia, ↓physical activity, which were all reduced by HRF inhibitors (N19, HRF-2CA)	**Egg allergy:** ↑serum HRF-reactive IgE, which was reduced by successful OIT[1]
Chronic idiopathic urticaria (CIU)		↑serum HRF, ↑serum HRF-reactive IgE
Pulmonary arterial hypertension (PAH)		↑plasma and lung HRF associated with exosomes

Oral immunotherapy (OIT[1]). ↓, decreased; ↑, increased.

Sites within the HRF molecule led to the discovery of HRF sequence-based competitive inhibitors, N19 and H3 peptides, as well as a monomeric mutant HRF-2CA, all of which blocked HRF–Ig interactions without affecting intracellular functions of TCTP. Administration of these inhibitors drastically reduced type 2 inflammation in mast cell-dependent murine models of atopic asthma and immediate hypersensitivity of the skin. Intranasal administration of recombinant HRF into naïve mice caused lung inflammation in an FcεRI and mast cell-dependent manner [24]. Thus, this study in 2012 solved several major questions about HRF, including the aforementioned issues (i) and (ii). More recently, Ando et al. showed that HRF dimers, but not monomers, are able to activate HRF-reactive IgE-bound mast cells and basophils [30]. Intragastric administration of HRF inhibitors, which preferentially targeted mast cells in the small intestine, strongly reduced diarrhea occurrence, intestinal inflammation, and systemic anaphylaxis in a murine model of food allergy [30,43]. Levels of HRF oligomers (including dimers) in the small intestine and HRF-reactive IgE in serum were increased in food allergic mice, but HRF oligomers were decreased by HRF inhibitors. Patients with egg allergy also had higher blood levels of HRF-reactive IgE, and successful oral immunotherapy led to reduced HRF-reactive IgE. Thus, these data suggest that in allergen-sensitized mice, secreted HRF oligomers bind to the Fab portion of IgE and reduce the threshold of allergen concentrations required to crosslink IgE-bound FcεRI to activate intestinal mast cells and basophils to elicit the food allergy phenotype (Figure 2).

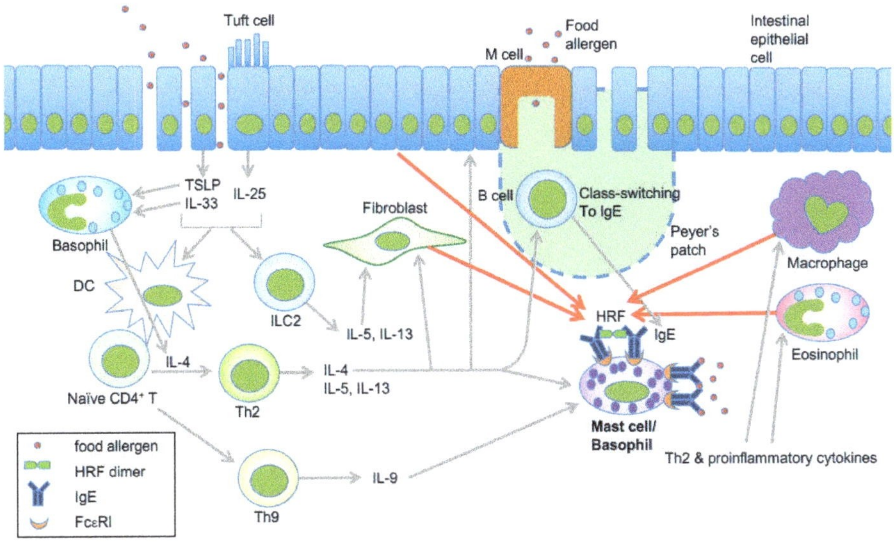

Figure 2. Model of HRF-mediated amplification of type 2 inflammation in food allergy. Epithelial damage or inflammation in the gut promotes increased entry of food allergens and secretion of the epithelial cytokines TSLP, IL-25, and IL-33 [44]. These cytokines initiate a Th2-skewed immune response. TSLP can enhance OX40L expression in dendritic cells, which induce Th2 cell differentiation of naïve CD4$^+$ T cells [45]. IL-25 secreted by tuft cells may help the expansion of type 2 innate lymphoid cells (ILC2) [46]. Th2 cells along with ILC2 cells promote the Th2 cell-mediated immune response, which includes IgE class switch recombination in B cells, eosinophil accumulation, and mastocytosis. IL-9 promotes the expansion of IL-9-producing mucosal mast cells [47]. Basophils are also required for production of antigen-specific IgE as well as oral allergen-induced food allergy during sensitization [48,49] and allergen challenge phases [50]. IL-4 derived from basophils stimulated by cytokines such as IL-33 seems to be required for Th2 cell differentiation [51], and IL-4 promotes intestinal mast cell accumulation and activation [52]. HRF dimer/oligomers secreted from several types of cells amplify intestinal inflammation by enhancing antigen/IgE-mediated activation of mast cells and basophils [30]. This is likely due to increased HRF secretion by several types of cells in response to Th2, proinflammatory and epithelial cytokines. Modified from ref. 66 with permission from the journal *Allergy*.

Another interesting drug candidate is a 7-mer peptide, called dTBP2. It was identified by phage display as a peptide more strongly bound to HRF dimer than to monomeric HRF [53]. dTBP2 ameliorated ovalbumin-induced airway inflammation in mice and reduced IL-8 release from BEAS-2B human bronchial epithelial cells. Recently, dehydrocostus lactone, a sesquiterpene from *Saussurea lappa Clarke*, which is able to bind to HRF dimers, was reported to suppress ovalbumin-induced airway inflammation [54]. However, given its action on various biological activities, it is premature to conclude that the anti-inflammatory effects of this compound are due to the inhibition of HRF dimer.

Atopic dermatitis is a heterogeneous disease in terms of the pathogenic role of the IgE–FcεRI axis [55,56]. Interestingly, atopic dermatitis patients have increased levels of HRF, and some patients have higher levels of HRF-reactive IgE compared to healthy individuals [57]. Polyclonal IgE molecules present in sera from atopic dermatitis patients activated mast cells [58], similar to highly cytokinergic IgE [59]. Topical administration of dTBP2 reduced allergen-induced atopic dermatitis in NC/Nga mice [60], a murine model of atopic dermatitis [56].

Chronic idiopathic urticaria (CIU) or chronic spontaneous urticaria is a disease of itchy red skin or skin colored hives with no known cause lasting for six weeks or more. IgG autoantibodies against IgE or FcεRI may contribute to CIU pathogenesis in 30%–40% of the patients [61]. Activation of skin mast cells plays a key role in skin inflammation of CIU. Interestingly, a recent study reported increased serum levels of both HRF and HRF-reactive IgE in CIU patients compared to healthy cohorts, and there was a linear correlation between HRF and HRF-reactive IgE concentrations in CIU patients [62]. Furthermore, the HRF-reactive IgE level was correlated with disease severity. The authors observed degranulation in the human mast cell line LAD-2 sensitized with serum of a CIU patient and stimulated with HRF. They suggested that synergistic actions of HRF and HRF-reactive IgE may play an important role in the CIU pathogenesis.

Pulmonary arterial hypertension (PAH) is a rare, but often lethal disease characterized by a sustained increase in pulmonary arterial pressure and severe vascular remodeling. Heritable PAH commonly involves mutations in bone morphogenetic protein receptor type II (*BMPR2*). Excessive proliferation of pulmonary vascular endothelial cells is seen in this disease caused by an imbalance between cell proliferation and apoptosis. Increased plasma and lung levels of HRF associated with exosomes derived from endothelial cells were found in PAH patients compared to normal subjects [63,64]. The exosome-derived HRF was taken up by pulmonary artery smooth muscle cells in in vitro co-cultures, and promoted proliferation and suppressed apoptosis of the latter cells [20,63]. These results suggest that HRF may not require a specific cell surface receptor for this type of intercellular communication, as extracellular HRF that has reached the interior of recipient cells would interact with its target molecules, potentially including Bcl-XL and Mcl-1. Interestingly, essentially all exosome-associated (and microparticle-associated) HRF in endothelial cells was dimeric [63]. However, there is no evidence that the function of intracellular TCTP molecules is operated by the dimeric form, as the vast majority of intracellular TCTP molecules is monomeric [30]. No definitive studies have been conducted to assign the functions of HRF/TCTP to either its monomeric or dimeric forms (or other forms) in PAH and other diseases.

5. Concluding Remarks

It is not easy to assign a particular pathogenic role to the secreted HRF molecules separate from the intracellular TCTP molecules. Targeting HRF is a promising approach toward prevention of allergic diseases such as food allergy and asthma [24,30,65]. However, all of the current HRF inhibitors have yet to be fully characterized as therapeutic agents. It is highly desirable to gain both pharmacological and genetic evidence before the field moves to clinical trials of candidate HRF inhibitors. However, genetic studies without affecting the function of intracellular TCTP are difficult if an experiment is conducted with TCTP conditional knockout (CKO) mice, including inducible CKO mice [40,66]. It is likely that the targeted cells may die because of their dependence of survival on TCTP. With such limitations, RNA interference (siRNA or shRNA) may be better suited to in vitro and in vivo experiments [66]. An alternative approach is to use heterozygous TCTP KO mice. Indeed, Pinkaew et al. showed that atherosclerotic lesions in $TCTP^{+/-}Ldlr^{-/-}Apobec1^{-/-}$ mice contain fewer macrophages and more apoptotic cells compared to $TCTP^{+/+}Ldlr^{-/-}Apobec1^{-/-}$ mice [67]. Transgenic overexpression may also be useful for analysis of HRF. Yeh et al. generated an inducible transgenic mouse model with HRF targeted to lung epithelial Clara cells [68]. They showed that HRF exacerbates the allergic asthmatic responses, although it is not clear whether secreted HRF was responsible for the worsened phenotype. Despite these obstacles, HRF inhibitors may be a promising approach toward preventing or treating food allergy and other IgE/HRF-dependent allergic diseases.

Author Contributions: Y.K. and K.K. drafted the manuscript. T.K. and Y.K. finished it.

Funding: This study was supported in part by grants from the US National Institutes of Health (HL124283, AI124734 and AI146042).

Conflicts of Interest: The authors declare no conflict of interest.

References

1. Metzger, H. Transmembrane signaling: The joy of aggregation. *J. Immunol.* **1992**, *149*, 1477–1487. [PubMed]

2. Mukai, K.; Tsai, M.; Saito, H.; Galli, S.J. Mast cells as sources of cytokines, chemokines, and growth factors. *Immunol. Rev.* **2018**, *282*, 121–150. [CrossRef] [PubMed]
3. Galli, S.J.; Tsai, M. IgE and mast cells in allergic disease. *Nat. Med.* **2012**, *18*, 693–704. [CrossRef] [PubMed]
4. MacDonald, S.M.; Lichtenstein, L.M. Histamine-releasing factors and heterogeneity of IgE. *Springer Semin. Immunopathol.* **1990**, *12*, 415–428. [CrossRef]
5. Kuna, P.; Reddigari, S.R.; Rucinski, D.; Oppenheim, J.J.; Kaplan, A.P. Monocyte chemotactic and activating factor is a potent histamine-releasing factor for human basophils. *J. Exp. Med.* **1992**, *175*, 489–493. [CrossRef]
6. Kuna, P.; Reddigari, S.R.; Schall, T.J.; Rucinski, D.; Sadick, M.; Kaplan, A.P. Characterization of the human basophil response to cytokines, growth factors, and histamine releasing factors of the intercrine/chemokine family. *J. Immunol.* **1993**, *150*, 1932–1943.
7. Dahinden, C.A.; Geiser, T.; Brunner, T.; von Tscharner, V.; Caput, D.; Ferrara, P.; Minty, A.; Baggiolini, M. Monocyte chemotactic protein 3 is a most effective basophil- and eosinophil-activating chemokine. *J. Exp. Med.* **1994**, *179*, 751–756. [CrossRef]
8. MacDonald, S.M.; Rafnar, T.; Langdon, J.; Lichtenstein, L.M. Molecular identification of an IgE-dependent histamine-releasing factor. *Science* **1995**, *269*, 688–690. [CrossRef]
9. Bommer, U.A.; Thiele, B.J. The translationally controlled tumour protein (TCTP). *Int. J. Biochem. Cell Biol.* **2004**, *36*, 379–385. [CrossRef]
10. Bommer, U.A. Cellular function and regulation of the translationally controlled tumor protein TCTP. *Open Allergy J.* **2012**, *5*, 19–32. [CrossRef]
11. Amson, R.; Pece, S.; Marine, J.C.; Di Fiore, P.P.; Telerman, A. TPT1/TCTP-regulated pathways in phenotypic reprogramming. *Trends Cell Biol.* **2013**, *23*, 37–46. [CrossRef] [PubMed]
12. Pinkaew, D.; Fujise, K. Fortilin: A potential target for the prevention and treatment of human diseases. *Adv. Clin. Chem.* **2017**, *82*, 265–300. [PubMed]
13. Choi, K.W.; Hsu, Y.C. To cease or to proliferate: New insights into TCTP function from a Drosophila study. *Cell Adh. Migr.* **2007**, *1*, 129–130. [CrossRef] [PubMed]
14. Koziol, M.J.; Gurdon, J.B. TCTP in development and cancer. *Biochem. Res. Int.* **2012**, *2012*. [CrossRef] [PubMed]
15. Schroeder, J.T.; Lichtenstein, L.M.; MacDonald, S.M. An immunoglobulin E-dependent recombinant histamine-releasing factor induces interleukin-4 secretion from human basophils. *J. Exp. Med.* **1996**, *183*, 1265–1270. [CrossRef]
16. Schroeder, J.T.; Lichtenstein, L.M.; MacDonald, S.M. Recombinant histamine-releasing factor enhances IgE-dependent IL-4 and IL-13 secretion by human basophils. *J. Immunol.* **1997**, *159*, 447–452.
17. Amzallag, N.; Passer, B.J.; Allanic, D.; Segura, E.; Thery, C.; Goud, B.; Amson, R.; Telerman, A. TSAP6 facilitates the secretion of translationally controlled tumor protein/histamine-releasing factor via a nonclassical pathway. *J. Biol. Chem.* **2004**, *279*, 46104–46112. [CrossRef]
18. Lespagnol, A.; Duflaut, D.; Beekman, C.; Blanc, L.; Fiucci, G.; Marine, J.C.; Vidal, M.; Amson, R.; Telerman, A. Exosome secretion, including the DNA damage-induced p53-dependent secretory pathway, is severely compromised in TSAP6/Steap3-null mice. *Cell Death Differ.* **2008**, *15*, 1723–1733. [CrossRef]
19. Yu, X.; Harris, S.L.; Levine, A.J. The regulation of exosome secretion: A novel function of the p53 protein. *Cancer Res.* **2006**, *66*, 4795–4801. [CrossRef]
20. Sirois, I.; Raymond, M.A.; Brassard, N.; Cailhier, J.F.; Fedjaev, M.; Hamelin, K.; Londono, I.; Bendayan, M.; Pshezhetsky, A.V.; Hebert, M.J. Caspase-3-dependent export of TCTP: A novel pathway for antiapoptotic intercellular communication. *Cell Death Differ.* **2011**, *18*, 549–562. [CrossRef]
21. MacDonald, S.M.; Lichtenstein, L.M.; Proud, D.; Plaut, M.; Naclerio, R.M.; MacGlashan, D.W.; Kagey-Sobotka, A. Studies of IgE-dependent histamine releasing factors: Heterogeneity of IgE. *J. Immunol.* **1987**, *139*, 506–512. [PubMed]
22. Zhang, K.; Max, E.E.; Cheah, H.K.; Saxon, A. Complex alternative RNA splicing of epsilon-immunoglobulin transcripts produces mRNAs encoding four potential secreted protein isoforms. *J. Biol. Chem.* **1994**, *269*, 456–462. [PubMed]
23. Wantke, F.; MacGlashan, D.W.; Langdon, J.M.; MacDonald, S.M. The human recombinant histamine releasing factor: Functional evidence that it does not bind to the IgE molecule. *J. Allergy Clin. Immunol.* **1999**, *103*, 642–648. [CrossRef]

24. Kashiwakura, J.; Ando, T.; Matsumoto, K.; Kimura, M.; Kitaura, J.; Matho, M.H.; Zajonc, D.M.; Ozeki, T.; Ra, C.; Macdonald, S.M.; et al. Histamine-releasing factor has a proinflammatory role in mouse models of asthma and allergy. *J. Clin. Investig.* **2012**, *122*, 218–228. [CrossRef] [PubMed]
25. Budde, I.K.; Aalbers, M.; Aalberse, R.C.; van der Zee, J.S.; Knol, E.F. Reactivity to IgE-dependent histamine-releasing factor is due to monomeric IgE. *Allergy* **2000**, *55*, 653–657. [CrossRef]
26. Kleine-Tebbe, J.; Kagey-Sobotka, A.; MacGlashan, D.W., Jr.; Lichtenstein, L.M.; MacDonald, S.M. Lectins do not distinguish between heterogenous IgE molecules as defined by differential activity of an IgE-dependent histamine releasing factor. *J. Allergy Clin. Immunol.* **1996**, *98*, 181–188. [CrossRef]
27. Shade, K.T.; Platzer, B.; Washburn, N.; Mani, V.; Bartsch, Y.C.; Conroy, M.; Pagan, J.D.; Bosques, C.; Mempel, T.R.; Fiebiger, E.; et al. A single glycan on IgE is indispensable for initiation of anaphylaxis. *J. Exp. Med.* **2015**, *212*, 457–467. [CrossRef]
28. Dore, K.A.; Kashiwakura, J.I.; McDonnell, J.M.; Gould, H.J.; Kawakami, T.; Sutton, B.J.; Davies, A.M. Crystal structures of murine and human Histamine-Releasing Factor (HRF/TCTP) and a model for HRF dimerisation in mast cell activation. *Mol. Immunol.* **2018**, *93*, 216–222. [CrossRef]
29. Kim, M.; Min, H.J.; Won, H.Y.; Park, H.; Lee, J.C.; Park, H.W.; Chung, J.; Hwang, E.S.; Lee, K. Dimerization of translationally controlled tumor protein is essential for its cytokine-like activity. *PLoS ONE* **2009**, *4*, e6464. [CrossRef]
30. Ando, T.; Kashiwakura, J.I.; Itoh-Nagato, N.; Yamashita, H.; Baba, M.; Kawakami, Y.; Tsai, S.H.; Inagaki, N.; Takeda, K.; Iwata, T.; et al. Histamine-releasing factor enhances food allergy. *J. Clin. Investig.* **2017**, *127*, 4541–4553. [CrossRef]
31. Yu, W.; Freeland, D.M.; Nadeau, K.C. Food allergy: Immune mechanisms, diagnosis and immunotherapy. *Nat. Rev. Immunol.* **2016**, *16*, 751–765. [CrossRef] [PubMed]
32. Fahy, J.V. Type 2 inflammation in asthma–present in most, absent in many. *Nat. Rev. Immunol.* **2015**, *15*, 57–65. [CrossRef] [PubMed]
33. Weidinger, S.; Beck, L.A.; Bieber, T.; Kabashima, K.; Irvine, A.D. Atopic dermatitis. *Nat. Rev. Dis Primers* **2018**, *4*, 1. [CrossRef] [PubMed]
34. Berin, M.C.; Mayer, L. Immunophysiology of experimental food allergy. *Mucosal Immunol.* **2009**, *2*, 24–32. [CrossRef] [PubMed]
35. Chinthrajah, R.S.; Hernandez, J.D.; Boyd, S.D.; Galli, S.J.; Nadeau, K.C. Molecular and cellular mechanisms of food allergy and food tolerance. *J. Allergy Clin. Immun.* **2016**, *137*, 984–997. [CrossRef]
36. Corazza, N.; Kaufmann, T. Novel insights into mechanisms of food allergy and allergic airway inflammation using experimental mouse models. *Allergy* **2012**, *67*, 1483–1490. [CrossRef]
37. Berin, M.C.; Mayer, L. Can we produce true tolerance in patients with food allergy? *J. Allergy Clin. Immun.* **2013**, *131*, 14–22. [CrossRef]
38. MacDonald, S.M. *Histamine Releasing Factors and IgE Heterogeneity*, 4th ed.; Mosby-Year Book Incorporated: St. Louis, MO, USA, 1993.
39. Sampson, H.A.; Broadbent, K.R.; Bernhisel-Broadbent, J. Spontaneous release of histamine from basophils and histamine-releasing factor in patients with atopic dermatitis and food hypersensitivity. *N. Engl. J. Med.* **1989**, *321*, 228–232. [CrossRef]
40. Chen, S.H.; Wu, P.S.; Chou, C.H.; Yan, Y.T.; Liu, H.; Weng, S.Y.; Yang-Yen, H.F. A knockout mouse approach reveals that TCTP functions as an essential factor for cell proliferation and survival in a tissue- or cell type-specific manner. *Mol. Biol. Cell* **2007**, *18*, 2525–2532. [CrossRef]
41. Susini, L.; Besse, S.; Duflaut, D.; Lespagnol, A.; Beekman, C.; Fiucci, G.; Atkinson, A.R.; Busso, D.; Poussin, P.; Marine, J.C.; et al. TCTP protects from apoptotic cell death by antagonizing bax function. *Cell Death Differ.* **2008**, *15*, 1211–1220. [CrossRef]
42. Koide, Y.; Kiyota, T.; Tonganunt, M.; Pinkaew, D.; Liu, Z.; Kato, Y.; Hutadilok-Towatana, N.; Phongdara, A.; Fujise, K. Embryonic lethality of fortilin-null mutant mice by BMP-pathway overactivation. *Biochim. Biophys. Acta* **2009**, *1790*, 326–338. [CrossRef] [PubMed]
43. Kawakami, Y.; Sielski, R.; Kawakami, T. Mouse Body Temperature Measurement Using Infrared Thermometer During Passive Systemic Anaphylaxis and Food Allergy Evaluation. *J. Vis. Exp.* **2018**, *139*, e58391. [CrossRef] [PubMed]

44. Leyva-Castillo, J.M.; Galand, C.; Kam, C.; Burton, O.; Gurish, M.; Musser, M.A.; Goldsmith, J.D.; Hait, E.; Nurko, S.; Brombacher, F.; et al. Mechanical skin injury promotes food anaphylaxis by driving intestinal mast cell expansion. *Immunity* **2019**, *50*, 1262–1275. [CrossRef] [PubMed]
45. Ito, T.; Wang, Y.H.; Duramad, O.; Hori, T.; Delespesse, G.J.; Watanabe, N.; Qin, F.X.; Yao, Z.; Cao, W.; Liu, Y.J. TSLP-activated dendritic cells induce an inflammatory T helper type 2 cell response through OX40 ligand. *J. Exp. Med.* **2005**, *202*, 1213–1223. [CrossRef] [PubMed]
46. Von Moltke, J.; Ji, M.; Liang, H.E.; Locksley, R.M. Tuft-cell-derived IL-25 regulates an intestinal ILC2-epithelial response circuit. *Nature* **2016**, *529*, 221. [CrossRef]
47. Chen, C.Y.; Lee, J.B.; Liu, B.; Ohta, S.; Wang, P.Y.; Kartashov, A.V.; Mugge, L.; Abonia, J.P.; Barski, A.; Izuhara, K.; et al. Induction of Interleukin-9-Producing Mucosal Mast Cells Promotes Susceptibility to IgE-Mediated Experimental Food Allergy. *Immunity* **2015**, *43*, 788–802. [CrossRef]
48. Muto, T.; Fukuoka, A.; Kabashima, K.; Ziegler, S.F.; Nakanishi, K.; Matsushita, K.; Yoshimoto, T. The role of basophils and proallergic cytokines, TSLP and IL-33, in cutaneously sensitized food allergy. *Int. Immunol.* **2014**, *26*, 539–549. [CrossRef]
49. Noti, M.; Kim, B.S.; Siracusa, M.C.; Rak, G.D.; Kubo, M.; Moghaddam, A.E.; Sattentau, Q.A.; Comeau, M.R.; Spergel, J.M.; Artis, D. Exposure to food allergens through inflamed skin promotes intestinal food allergy through the thymic stromal lymphopoietin-basophil axis. *J. Allergy Clin. Immunol.* **2014**, *133*, 1390–1399. [CrossRef]
50. Kashiwakura, J.I.; Ando, T.; Karasuyama, H.; Kubo, M.; Matsumoto, K.; Matsuda, T.; Kawakami, T. The basophil-IL-4-mast cell axis is required for food allergy. *Allergy* **2019**. [CrossRef]
51. Miyake, K.; Karasuyama, H. Emerging roles of basophils in allergic inflammation. *Allergol. Int.* **2017**, *66*, 382–391. [CrossRef]
52. Burton, O.T.; Darling, A.R.; Zhou, J.S.; Noval, M.-R.; Jones, T.G.; Gurish, M.F.; Chatila, T.A.; Oettgen, H.C. Direct effects of IL-4 on mast cells drive their intestinal expansion and increase susceptibility to anaphylaxis in a murine model of food allergy. *Mucosal Immunol.* **2013**, *6*, 740–750. [CrossRef] [PubMed]
53. Kim, M.; Chung, J.; Lee, C.; Jung, J.; Kwon, Y.; Lee, K. A peptide binding to dimerized translationally controlled tumor protein modulates allergic reactions. *J. Mol. Med. (Berl.)* **2011**, *89*, 603–610. [CrossRef] [PubMed]
54. Pyun, H.; Kang, U.; Seo, E.K.; Lee, K. Dehydrocostus lactone, a sesquiterpene from Saussurea lappa Clarke, suppresses allergic airway inflammation by binding to dimerized translationally controlled tumor protein. *Phytomedicine* **2018**, *43*, 46–54. [CrossRef] [PubMed]
55. Jin, H.; He, R.; Oyoshi, M.; Geha, R.S. Animal models of atopic dermatitis. *J. Investig. Dermatol.* **2009**, *129*, 31–40. [CrossRef] [PubMed]
56. Kawakami, T.; Ando, T.; Kimura, M.; Wilson, B.S.; Kawakami, Y. Mast cells in atopic dermatitis. *Curr. Opin. Immunol.* **2009**, *21*, 666–678. [CrossRef]
57. Kashiwakura, J.; Okayama, Y.; Furue, M.; Kabashima, K.; Shimada, S.; Ra, C.; Siraganian, R.P.; Kawakami, Y.; Kawakami, T. Most highly cytokinergic IgEs have polyreactivity to autoantigens. *Allergy Asthma Immunol. Res.* **2012**, *4*, 332–340. [CrossRef]
58. Kashiwakura, J.; Kawakami, Y.; Yuki, K.; Zajonc, D.M.; Hasegawa, S.; Tomimori, Y.; Caplan, B.; Saito, H.; Furue, M.; Oettgen, H.C.; et al. Polyclonal IgE induces mast cell survival and cytokine production. *Allergol. Int. Off. J. Jpn. Soc. Allergol.* **2009**, *58*, 411–419. [CrossRef]
59. Kitaura, J.; Song, J.; Tsai, M.; Asai, K.; Maeda-Yamamoto, M.; Mocsai, A.; Kawakami, Y.; Liu, F.T.; Lowell, C.A.; Barisas, B.G.; et al. Evidence that IgE molecules mediate a spectrum of effects on mast cell survival and activation via aggregation of the FcepsilonRI. *Proc. Natl. Acad. Sci. USA* **2003**, *100*, 12911–12916. [CrossRef]
60. Jin, X.H.; Lim, J.; Shin, D.H.; Maeng, J.; Lee, K. Dimerized Translationally Controlled Tumor Protein-Binding Peptide Ameliorates Atopic Dermatitis in NC/Nga Mice. *Int. J. Mol. Sci.* **2017**, *18*, 256. [CrossRef]
61. Goh, C.L.; Tan, K.T. Chronic autoimmune urticaria: Where we stand? *Indian J. Derm.* **2009**, *54*, 269–274. [CrossRef]
62. Huang, X.; Li, Z.; Sun, R. Synergistic Actions of Histamine-Releasing Factor and Histamine Releasing Factor-Reactive IgE in Chronic Urticaria. *Int. Arch. Allergy Immunol.* **2017**, *172*, 27–32. [CrossRef] [PubMed]

63. Ferrer, E.; Dunmore, B.J.; Hassan, D.; Ormiston, M.L.; Moore, S.; Deighton, J.; Long, L.; Yang, X.D.; Stewart, D.J.; Morrell, N.W. A Potential Role for Exosomal Translationally Controlled Tumor Protein Export in Vascular Remodeling in Pulmonary Arterial Hypertension. *Am. J. Respir Cell Mol. Biol.* **2018**, *59*, 467–478. [CrossRef] [PubMed]
64. Lavoie, J.R.; Ormiston, M.L.; Perez-Iratxeta, C.; Courtman, D.W.; Jiang, B.; Ferrer, E.; Caruso, P.; Southwood, M.; Foster, W.S.; Morrell, N.W.; et al. Proteomic analysis implicates translationally controlled tumor protein as a novel mediator of occlusive vascular remodeling in pulmonary arterial hypertension. *Circulation* **2014**, *129*, 2125–2135. [CrossRef] [PubMed]
65. Wills-Karp, M. Histamine-releasing factor: A promising therapeutic target for food allergy. *J. Clin. Investig.* **2017**, *127*, 4238–4241. [CrossRef] [PubMed]
66. Pinkaew, D.; Chattopadhyay, A.; King, M.D.; Chunhacha, P.; Liu, Z.; Stevenson, H.L.; Chen, Y.; Sinthujaroen, P.; McDougal, O.M.; Fujise, K. Fortilin binds IRE1alpha and prevents ER stress from signaling apoptotic cell death. *Nat. Commun.* **2017**, *8*, 18. [CrossRef]
67. Pinkaew, D.; Le, R.J.; Chen, Y.; Eltorky, M.; Teng, B.B.; Fujise, K. Fortilin reduces apoptosis in macrophages and promotes atherosclerosis. *Am. J. Physiol Heart Circ. Physiol* **2013**, *305*, H1519–H1529. [CrossRef]
68. Yeh, Y.C.; Xie, L.; Langdon, J.M.; Myers, A.C.; Oh, S.Y.; Zhu, Z.; Macdonald, S.M. The effects of overexpression of histamine releasing factor (HRF) in a transgenic mouse model. *PLoS ONE* **2010**, *5*, e11077. [CrossRef]

cells

MDPI

Article

TCTP from *Loxosceles Intermedia* (Brown Spider) Venom Contributes to the Allergic and Inflammatory Response of Cutaneous Loxoscelism

Marianna Boia-Ferreira [1], Kamila G. Moreno [1], Alana B. C. Basílio [1], Lucas P. da Silva [1], Larissa Vuitika [1], Bruna Soley [2], Ana Carolina M. Wille [3], Lucélia Donatti [1], Katia C. Barbaro [4], Olga M. Chaim [1,5], Luiza Helena Gremski [1], Silvio S. Veiga [1] and Andrea Senff-Ribeiro [1,*]

1 Department of Cell Biology, Federal University of Paraná, Curitiba 81531-980, PR, Brazil; marianna.boia@gmail.com (M.B.-F.); kamilamoreno7@gmail.com (K.G.M.); alana_abcb@hotmail.com (A.B.C.B.); spedrosa.lucas@gmail.com (L.P.d.S.); larissavuitika2@gmail.com (L.V.); donatti@ufpr.br (L.D.); olgachaim@ufpr.br or ochaim@ucsd.edu (O.M.C.); luiza_hg@yahoo.com.br (L.H.G.); veigass@ufpr.br (S.S.V.)

2 Department of Pharmacology, Federal University of Paraná, Curitiba 81531-980, PR, Brazil; brunasoley@gmail.com

3 Department of Structural and Molecular Biology, State University of Ponta Grossa, Ponta Grossa 84030-900, PR, Brazil; anacarolina.wille@yahoo.com.br

4 Laboratory of Immunopathology, Butantan Institute, São Paulo 05503-900, SP, Brazil; katiacbarbaro@hotmail.com

5 Department of Pharmacology, School of Medicine, University of California San Diego, La Jolla, CA 92093, USA

* Correspondence: senffribeiro@gmail.com or senffribeiro@ufpr.br; Tel.: +55-41-3361-1779

Received: 22 August 2019; Accepted: 10 November 2019; Published: 22 November 2019

Abstract: LiTCTP is a toxin from the Translationally Controlled Tumor Protein (TCTP) family identified in *Loxosceles* brown spider venoms. These proteins are known as histamine-releasing factors (HRF). TCTPs participate in allergic and anaphylactic reactions, which suggest their potential role as therapeutic targets. The histaminergic effect of TCTP is related to its pro-inflammatory functions. An initial characterization of LiTCTP in animal models showed that this toxin can increase the microvascular permeability of skin vessels and induce paw edema in a dose-dependent manner. We evaluated the role of LiTCTP in vitro and in vivo in the inflammatory and allergic aspects that undergo the biological responses observed in Loxoscelism, the clinical condition after an accident with *Loxosceles* spiders. Our results showed LiTCTP recombinant toxin (LiRecTCTP) as an essential synergistic factor for the dermonecrotic toxin actions (LiRecDT1, known as the main toxin in the pathophysiology of Loxoscelism), revealing its contribution to the exacerbated inflammatory response clinically observed in envenomated patients.

Keywords: *Loxosceles*; brown spider; TCTP; venom; toxin; HRF

1. Introduction

LiTCTP is a protein from the Translationally Controlled Tumor Protein (TCTP) family that was found in the *Loxosceles intermedia* brown spider venom, initially in a cDNA library of the *L. intermedia* venom gland and confirmed in the transcriptome analysis of the venom gland [1,2]. Although it represented only 0.4% of the toxin-related transcripts, it was positively identified by immunodetection in the whole venom of different *Loxosceles* species (*L. intermedia*, *L. gaucho*, and *L. laeta*) [3], which indicates its biological conservation and importance. TCTP proteins are known as histamine releasing factors (HRF) and activators of mast cells and basophils triggering the release of histamine [4,5]. Mast cells are intimately involved in allergic and anaphylactic reactions, and an increasing body of evidence involves these cells and their mediators in the pathophysiology of inflammation [6]. *Loxosceles* venoms are responsible for severe skin lesions at the bite site, characterized by intense inflammatory content, which can evolve to necrotic conditions [7,8]. Hypersensitivity or even allergic reactions are also reported as clinical features of Loxoscelism [9,10]. Mast cells were already mentioned as involved in biological responses evoked by *Loxosceles* venom toxins, as inflammatory response was partially reduced in compound 48/80-pretreated animals [11]. Previous study of LiRecTCTP, the recombinant isoform of LiTCTP, showed this toxin increases microvascular permeability of skin vessels, causing a diffuse pattern of dye leakage. Moreover, LiRecTCTP induced paw edema, which reached its maximum after 5 min of inoculation (an early effect compared to dermonecrosis) [2]. TCTP was already described as a putative therapeutical target in asthma and allergy due to its pro-inflammatory extracellular effects [5,12]. Herein, we studied the participation and effects of LiRecTCTP toxin in the biological histaminergic and inflammatory response observed in Loxoscelism. LiRecTCTP was also studied in combination with the well-known LiRecDT1 Brown spider toxin, a recombinant isoform of phospholipase-D (PLD) of *L. intermedia*. LiRecDT1 can cause the majority of whole venom-induced effects in vitro and in vivo. Meanwhile, LiRecTCTP was shown to enhance the biological response to LiRecDT1, pointing to its contribution to the exacerbated inflammatory response observed clinically in the patients. Our findings strengthen the idea of inhibition of LiTCTP and mast cell activation effects as promising therapeutic approach to reduce the inflammatory events responsible for the main symptoms in cutaneous Loxoscelism.

2. Materials and Methods

2.1. Reagents

The whole venom from *L. intermedia* was obtained by electrostimulation (15 V) of the cephalothorax of spiders, solubilized in PBS, and maintained frozen until use [13]. *L. intermedia* spiders were captured in the wild with the authorization of the Brazilian Governmental Agency "Instituto Chico Mendes de Conservação da Biodiversidade" (Number 29801-1). Ni^{2+}-NTA agarose was purchased from Invitrogen (Carlsbad, CA, USA). DMEM media were purchased from Gibco (Carlsbad, CA, USA). The molecular mass markers were acquired from Sigma Aldrich (St. Louis, MO, USA). Evans Blue dye was purchased from Vetec (São Paulo, Brazil). The Compound 48/80, cromolyn sodium salt (cromolyn), promethazine hydrochloride (promethazine), cimetidine hydrochloride (cimetidine), and thioperamide maleate salt (thioperamide) were purchased from Sigma Aldrich. Ketamine and Sedanew® (xylazin 10%) were from Agribands (Campinas, Brazil) and Univet (São Paulo, Brazil), respectively.

2.2. Recombinant Protein Expression

The pET-14b cDNA construct [2] was transformed into one-shot *E. coli* BL21(DE3) pLysS competent cells (Invitrogen), plated on LB agar medium containing ampicillin (100 µg/mL) (Sigma Aldrich), and chloramphenicol (34 µg/mL) (Sigma Aldrich). One colony was then incubated in 10 mL of LB broth (with antibiotics) and allowed to grow overnight at 37 °C under orbital agitation. Then, this pre-culture was expanded into 1L of LB broth with antibiotics and allowed to grow at 37 °C until the OD at 550 nm reached 0.5. For the induction of heterologous protein expression, isopropyl-d-thiogalactoside (IPTG, ThermoFisher Scientific, Waltham, MA, USA) was added at a final concentration of 0.1 mM, and induction of the culture was performed for 4 h at 23 °C.

2.3. Recombinant Protein Purification

LiRecTCTP was purified by affinity chromatography using Ni^{+2}-NTA column (Invitrogen) as described by Sade and colleagues [2], with modifications. Briefly, *Escherichia coli* cells were lysed by thaw–freeze cycles and disrupted by cycles of sonication. The cell lysate was centrifuged (20,000× *g*, 20 min, 4 °C), and the supernatant was incubated with 1 mL Ni^{2+}-NTA beads for 1 h at 4 °C.

The recombinant protein was washed with wash buffer (50 mM sodium phosphate pH 6.0, 500 mM NaCl, 35 mM imidazole, Sigma Aldrich), eluted with elution buffer (50 mM sodium phosphate pH 8.0, 500 mM NaCl, 350 mM imidazole), and analyzed by 12.5% SDS-PAGE under reducing conditions. Wash buffer pH was important to the astringent condition for protein binding to the column.

2.4. Protein Quantification and Gel Electrophoresis

The total protein concentration of different samples and, especially, the purified LiRecTCTP were determined by the Coomassie Blue method (BioRad, Hercules, CA, USA) [14]. For protein quality analysis, 12.5% SDS-PAGE (sodium dodecyl sulfate-polyacrylamide gel electrophoresis) was performed under reducing conditions.

2.5. Circular Dichroism Spectroscopy (CD)

Recombinant LiRecTCTP was dialyzed at 4 °C against a phosphate buffer (20 mM NaH_2PO_4/Na_2HPO_4 and 150 mM NaCl at pH 7.4) to a final concentration of 0.5 mg/mL. The UV-CD spectra were obtained in a Jasco J-810 spectropolarimeter (Jasco Corporation, Tokyo, Japan) using a 1 mm cuvette, as described previously [15]. The final spectrum (0.5 nm intervals) was the average of three measurements, performed at a rate of 50 nm/min, using a response time of 8 s and a bandwidth of 1 nm. The temperature was kept constant at 20 °C. The data units were expressed as molar ellipticity and plotted by GraphPad Prism 6 software. The results of deconvolution analyses and percentages of secondary structure were predicted by K2D3 tool (https://onlinelibrary.wiley.com/doi/abs/10.1002/prot.23188).

2.6. Cell Culture

RBL-2H3 cell line was obtained as a courtesy of Professor Maria Celia Jamur (University of São Paulo, School of Medicine, Ribeirão Preto, SP, Brazil) and were grown in DMEM medium supplemented with 10% FCS (Gibco) and antibiotics penicillin (10,000 U/mL) (Sigma Aldrich) and streptomycin (10 mg/mL) (Sigma Aldrich). RBL-2H3 cells are a rat basophilic leukemia cell line used as a mast-like cell model [16]. RBL-2H3 cells grow as a monolayer, were subcultured (using trypsin/EDTA 2 mM, Gibco) and used until the 10th passage. Cell cultures were kept at 37 °C in a humidified atmosphere with 5% CO_2.

2.7. MTT Assay

RBL-2H3 cells (5×10^4 cells per well) cells were plated in 96-well plates and then grown in medium containing 10% FCS at 37 °C in a humidified 5% CO_2 incubator. After 16 h, cells were washed with Tyrode's Buffer (137 mM NaCl, 2.7 mM KCl, 1mM $MgCl_2$, 1.8 mM $CaCl_2$, 0.2 mM Na_2HPO_4, 5.5 mM Glucose, 10 mM HEPES, and 0.1% BSA), and then incubated with LiRecTCTP (10, 50, and 100 µg/mL), total venom from *L. intermedia* (10, 50, and 100 µg/mL), 48/80 compound (100 µg/mL) (positive control for degranulation), and PBS or the recombinant toxin LiRecDT1H12A (100 µg/mL) (as negative controls). After 2 h, media was removed and replaced by MTT solution (0.5 mg/mL) (Sigma Aldrich). It is important to mention that incubation with toxins did not cause any detachment of cells from the plates. Cells were again incubated for 3 h at 37 °C. The MTT solution was removed, and formed formazan crystals of each sample were solubilized with DMSO (100 µL) (Sigma Aldrich). The dehydrogenases activity for cell viability assessment was quantified spectrometrically in 550 nm. MTT assay was performed in pentaplicate, and the results are shown as mean ± s.d. of three independent experiments.

2.8. In vitro Mast Cell Degranulation Induced by LiRecTCTP

The release of granular beta-hexosaminidase enzyme was measured in the supernatants obtained from RBL-2H3 rat cell line exposed to the recombinant toxins. For this, 5×10^4 cells were plated in medium with 10% FCS. After 16 h, cells were washed, and the medium was replaced by Tyrode's buffer containing LiRecTCTP (10, 50, and 100 μg/mL) with or without cromolyn (10 μM), total venom of *L. intermedia* (10, 50, and 100 μg/mL), 48/80 compound (100 μg/mL) (positive control), and PBS or the recombinant toxin LiRecDT1H12A (100 μg/mL) (as negative controls) for 2 h at 37 °C in a humidified 5% CO_2 incubator. From each experimental sample to be quantified, five aliquots (10 μL) of the supernatants were taken as pentaplicates to another microwell plate. RBL-2H3 cells incubated only with the Tyrode's (TGB) buffer were lysed with 0.5% Triton X-100 (200 μL) (Sigma Aldrich) to evaluate the total enzyme content as 100% reference. To all replicates, 90 μL of the substrate solution containing 1.3 mg/mL of p-nitrophenyl-*N*-acethyl-b-D-glucosamine (Sigma Aldrich) in 0.1 M citrate solution (pH 4.5) were added, and plates incubated for 30 min at 37 °C. Reactions were stopped by addition of 100 μL of 0.2 M glycine solution (pH 10.7), and OD at 405 nm was determined. The extent of secretion was expressed as the percentage of the total beta-hexosaminidase activity in the wells discounted of the values obtained in the supernatant of unstimulated cells [17]. Beta-hexosaminidase activity results are shown as mean ± s.d. of three independent experiments.

2.9. *Scanning Electronic Microscopy (SEM)*

Previously plated on glass coverslips, RBL-2H3 cells were exposed to PBS (as negative control) and LiRecTCTP (100 and 200 μg/mL) in Tyrode's Buffer (TGB) for 2 h at 37 °C. After, cells were washed with TGB and fixed with Karnovsky (2% formaldehyde, 2.5% glutaraldehyde in 0.1 M of sodium cacodylate buffer pH 7.2 at 4 °C) [18]. Then, fixed cells were dehydrated, and critical-point drying was performed using a Balzers CPD-010 (Balzers Instruments, Balzers, Liechtenstein) with carbonic gas. Metallization in gold was performed using a Balzers SCD-030 (Balzers Instruments). The samples were observed and photographed with a JEOL-JSM 6360 LV scanning electron microscope (JEOL Ltd., Tokyo, Japan) at the Electron Microscopy Center, Federal University of Paraná (Curitiba, PR, Brazil).

2.10. *Calcium Influx Assay*

Calcium influx was measured as previously described [19,20]. Briefly, cultured RBL-2H3 cells were removed and washed with PBS. After, cells were loaded with Fluo-4 AM (10 μM) (ThermoFisher) in buffer with Pluronic F-127 (0.01%) for 30 min at 37 °C. Subsequently, cells were washed twice with Tyrode's (TGB) buffer without calcium and equilibrated for 30 min at room temperature. Then, 5×10^5 cells/well were incubated in 96 wells black plates with PBS (negative control), LiRecTCTP (50 and 100 μg/mL), and LiRecTCTP (100 μg/mL) combined with cromolyn (10 μM) for 5, 15, 30, 60, and 90 min. The resulting fluorescence was quantified in Tecan Infinite M200 spectrofluorometer (Tecan, Männedorf, Switzerland) using an excitation wavelength of 485 nm and measuring emission at 535 nm. Calcium influx assay was performed in triplicate, and the results are shown as mean ± s.d. of three independent experiments.

2.11. *Quantitative PCR*

Total RNA was extracted from cells using TRIzol Reagent (Life Technologies, Carlsbad, CA, USA) according to the manufacturer's instructions. RNA was converted in cDNA using High-Capacity RNA-to-cDNA™ kit (Applied Biosystems, Foster City, CA, USA). The cDNA concentrations were measured with a NanoVue Plus spectrophotometer (GE Healthcare, Chicago, IL, USA) and 50 ng cDNA was used. Real-time quantitative PCR was performed using the Power SYBR Green PCR Master Mix in Step One Plus Real-Time PCR System (Applied Biosystems). Primers for IL-3 (sense 5′-ACAATGGTTCTTGCCAGCTCTAC-3′ antisense 5′-AGGAGCGGGAGCAGCAT-3′), IL-4 (sense 5′-CAGGGTGCTTCGCAAATTTTAC-3′ anti-sense 5′-ACCGAGAACCCCAGACTTGTT-3′), IL-13 (sense 5′-GCTCTCGCTTCGCTTGGTGGTC-3′ anti-sense 5′CATCCGAGGCCTTTTGGTTACAG-3′), and GAPDH (sense 5′-TTCACCACCATGGAGAAGGC-3′ antisense 5′- GGCATGGACTGTGGTCATGA-3′) were based on validated sequences from Primer Bank [21]. GAPDH mRNA was used to normalize data, with fold change calculated by the comparative Ct method (ΔΔCT method), as previously described [22]. Results are shown as mean ± s.d. of three independent experiments.

2.12. Animals

Adult Swiss mice (25–30 g) from the Central Animal House of the Federal University of Paraná and adult rabbits (3 kg) from CPPI (Centro de Pesquisa e Produção de Imunobiológicos, Piraquara, Brasil) were used for in vivo experiments with the whole venom of *Loxosceles intermedia* and/or the recombinant toxins (LiRecTCTP, LiRecDT1, and GFP). All procedures involving animals were carried out in accordance with "Brazilian Federal Laws," following the Institutional Ethics Committee for Animal Studies Guidelines from Federal University of Paraná, which approved the project methodology concerning animal studies (Approval Certificates Numbers 730 and 1183).

2.13. Pharmacological Treatments (In Vivo)

The following pretreatments were applied into mice in order to investigate LiRecTCTP effects on edema formation and vascular permeability in vivo: cromolyn (mast cell degranulation inhibitor), 30 mg/kg, administered i.p. in three consecutive days before exposure to LiRecTCTP; promethazine (histamine type 1 receptor (H1R) antagonist), 5 mg/kg, administered i.v. 30 min before exposure to LiRecTCTP; cimetidine (histamine type 2 receptor (H2R) antagonist), 15 mg/kg, administered i.p. 2 h before exposure to LiRecTCTP; thioperamide (histamine type 3 and 4 receptor (H3/H4R) antagonist), 20 mg/kg, administered i.p. 30 min before exposure to LiRecTCTP [23]. Sterile PBS was used as a negative control for pharmacological treatments and LiRecTCTP injections.

2.14. Effects on Vascular Permeability

Changes in vascular permeability were assessed by visualizing extravasation of Evans Blue dye into the extravascular compartment of the skin [2,24,25]. Briefly, groups of five mice (per condition) were treated with the different types of inhibitors (as previously described in Pharmacological Treatments item). A dilution of the dye in PBS solution (30 mg/kg) was injected intravenously (100 μL) prior to intradermal injection of LiRecTCTP (10 μg) or PBS (negative control) (50 μL). After 60 min, animals were euthanized (intraperitoneal injection of ketamine 30 mg/kg and xylazin 5 mg/kg), and dorsal skin was removed for visualization of dye leakage and photographed. The patches of skin were excised and incubated in 2 mL of formamide at room temperature for five days, after which the absorbance of the resulting supernatant was measured at 595 nm. Results are of one representative experiment from three independent biological replicates (data shown in Figure 4). Alternatively, groups of five mice (per condition) were injected intravenously (100 μL) with a dilution of Evans Blue in PBS solution (30 mg/kg) prior to intradermal injection of LiRecTCTP (5 μg and 10 μg), LiRecDT1 (1 μg), recombinant GFP (10 μg) (negative control), and the same volume of PBS (50 μL) (negative control). After 60 min, animals were euthanized (as described above), and dorsal skin was removed for visualization of dye leakage and photographed (data shown in Figure 5).

2.15. Paw Edema-Forming Activity

Paw edema development was measured at different time intervals as previously performed [2,26]. Briefly, groups of five mice (per condition) were treated with different types of inhibitors (as previously described in Pharmacological Treatments item) and were injected subcutaneously into the right hind paw with LiRecTCTP (10 μg dissolved in sterile PBS). Negative control mice were injected with the same volume of PBS (30 μL). Edema was evaluated by examining changes in paw thickness using a calibrated digital micrometer (Digimess, São Paulo, SP, Brazil) at the following time points: immediately after subcutaneous injection (t zero), 5, 10, 20, 30, 60, 120, 240, 360, and 720 min after injection. Results are shown as mean ± s.e.m of one representative experiment from three independent biological replicates.

2.16. Dermonecrosis In Vivo

For assessment of dermonecrotic effects, 10 and 20 μg of LiRecTCTP, 1 μg of LiRecDT1 (wild type dermonecrotic toxin), and 20 μg of GFP (a recombinant protein without relevant biological activity) [2,27] were injected subcutaneously into a shaved area of the rabbit dorsal skin. Animals were observed over the course of dermonecrotic lesion progression. Animal skin was photographed immediately after injection and after 24 h of toxins application. The same rabbit received all the 7 samples (divided in the both dorsal sides of the animal) (PBS, GFP, LiRecDT1, LiRecTCTP 10 μg, LiRecTCTP 20 μg, LiRecDT1+LiRecTCTP 10 μg, and LiRecDT1+ LiRecTCTP 20 μg). Experiments were initially performed in two animals and then repeated using four animals. Images show a photograph from the skin of a representative rabbit. Animals were euthanized using intramuscular injection of ketamine (240 mg/kg) and xylazin (27 mg/kg). After euthanasia of animals, skin samples were harvested for histopathological analysis and myeloperoxidase (MPO) activity assay.

2.17. Histological Methods for Light Microscopy

Rabbit skin pieces from animals, which were previously subcutaneously inoculated with recombinant toxins, were collected. The tissue samples were fixed in "ALFAC" (ethanol 85%, formaldehyde 10%, and glacial acetic acid 5%) for 16 h at room temperature. After fixation, samples were dehydrated in a graded series of ethanol before paraffin embedding (for 2 h at 58 °C). Then, thin tissue sections (4 μm) were processed and stained with hematoxylin and eosin (H & E). Histological sections were analyzed in Image J analysis software (v.1.x) to quantify edema formation by the measurement of the area between the epidermis and adipose tissue, histological section area observed after GFP protein inoculation was considered the control for comparisons.

2.18. Myeloperoxidase (MPO) Activity Assay

The activity of tissue myeloperoxidase (MPO) in rabbit skin was evaluated 24 h after subcutaneous injection of toxins in the animals as previously described [27]. Briefly, a 6 mm skin tissue punch (biopsy) was placed in 0.75 mL of 80 mM sodium-phosphate buffer, pH 5.4, containing 0.5% hexadecyltrimethylammonium bromide (HTAB) (Sigma Aldrich), and then homogenized (45 s at 0 °C) in a motor-driven homogenizer. The homogenates were decanted into a microfuge tubes, and the vessel was washed with 0.75 mL of HTAB-buffer. The wash was added to tube, and the 1.5 mL sample was centrifuged at 12,000× g at 4 °C for 15 min. Samples in triplicate (30 μL) of the resulting supernatant were added into 96-well microtiter plates. For the assay, 200 μL of a mixture containing 100 μL of 80 mM PBS (pH 5.4), 85 μL of 0.22 M PBS (pH 5.4), and 30 μL of 0.017% hydrogen peroxide (w/w) were added to the wells. The reaction was started with the addition of 20 μL 18.4 mM TMB dihydrochloride (Sigma Aldrich) in dimethylformamide. Plates were incubated at 37 °C for 10 min, and then the reactions were stopped by the addition of 30 μL of 1.46 M sodium acetate, pH 3.0. Enzyme activity was determined colorimetrically using a plate reader set to measure absorbance at 630 nm and was expressed as mOD/biopsy. Results are shown as mean ± s.e.m of three independent biological replicates.

2.19. Statistical Analysis

Statistical analysis of MTT, Quantitative PCR, Calcium Influx, and Mast Cell Degranulation (beta-hexosaminidase) assays were performed using one-way ANOVA post-hoc Tukey test for average comparisons using the GraphPad Prism 6 program (GraphPad Software, San Diego, CA, USA). Statistical significance was established at $p < 0.1$. Statistical analysis of Paw Edema in vivo assay was performed using two-way ANOVA post-hoc Tukey test for average comparisons using the GraphPad Prism 6 program. Statistical significance was established at $p < 0.1$. Statistical analysis of Myeloperoxidase Activity in vivo was performed using Student's T test for average comparisons using the GraphPad Prism 6 program. Statistical significance was established at $p < 0.1$.

3. Results

3.1. LiRecTCTP Expression and Purification

LiRecTCTP expression was performed using the same heterologous system previously described, using *E. coli* and a His-tag [2], but we used an improved protocol of purification. In the former protocol, recombinant toxin was purified in native conditions in a 2-step chromatographic approach: Ni+2-NTA affinity chromatography using, and ion-exchange chromatography using DEAE agarose [2]. Herein, LiRecTCTP was purified under native conditions in just one step of chromatography (Ni^{+2}-affinity chromatography). This new protocol resulted in highly purified recombinant toxin and the yield was 16 mg/L of *E. coli* culture (Figure 1A). Purified LiRecTCTP toxin was submitted to circular dichroism to analyze protein folding. Deconvolution results show 43% of defined secondary structures as alpha-helix and beta-sheets (Figure 1B,C). These results are in agreement with previous data obtained for LiRecTCTP [2].

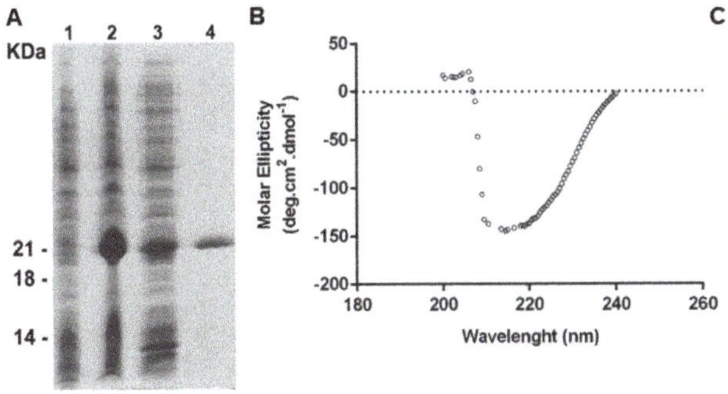

C

CD DECONVOLUTION (%)			
PROTEIN	α- HELIX	β- STRAND	OTHERS
LiRecTCTP	9%	34%	57%

Figure 1. Heterologous expression, purification, and circular dichroism spectroscopy CD analyses of recombinant LiTCTP. (**A**) SDS-PAGE (12.5%) analysis of recombinant LiTCTP toxin expression stained with Coomassie blue dye. Lane 1 show *E. coli* BL21 (DE3) pLysS cells before induction with IPTG. Lane 2 shows *E. coli* BL21(DE3) pLysS after induction for 4h with 0.1 mM isopropyl-d-thiogalactoside (IPTG) (supernatant from cell lysates obtained by freezing and thawing in extraction buffer before incubation with Ni^{2+}-NTA beads). Lane 3 depicts the void from Ni^{2+}-NTA chromatography. Lane 4 shows eluted recombinant protein from Ni^{2+}-NTA beads. Molecular mass markers are shown on the left. (**B**) The UV-CD spectrum was obtained in a Jasco J-810 spectropolarimeter (Jasco Corporation, Tokyo, Japan) by diluting the sample at 0.5 mg/mL in phosphate buffer, pH 7.4 at 20 °C. Graphic representation was plotted using GraphPad Prism 6. (**C**) The deconvolution of data, α-helix and β-sheet percentages of LiRecTCTP structure was calculated using K2D3 tool.

3.2. LiRecTCTP Activity on RBL-2H3 Cells

LiRecTCTP activity on mast cells degranulation was evaluated using RBL-2H3, a mast cell–like cell line, originally a rat basophilic leukemia cell. Initially, a cytotoxic effect of LiRecTCTP (100 μg/mL) on these cells was disregarded by evaluating cells viability (MTT assay) and morphology (SEM) (Figure 2A,C) after 2h-treatment with LiRecTCTP (10, 50, and 100 μg/mL). LiRecDT1H12A, a mutated and almost inactive toxin (only residual levels of activity) produced in the same way (heterologous system and chromatographic purification protocols), was included in the experiment (as negative control), as well as a compound that triggers degranulation (48/80, a positive control). Crude venom activity was also evaluated by MTT assay, and the resulting absorbance did not differ from controls. We did not observe deleterious effects of LiRecTCTP (100 μg/mL) in the viability/activity measured by MTT metabolization in formazan salt (Figure 2A) or in the cellular morphology shown in SEM (Figure 2C). Only 200 μg/mL of LiRecTCTP induced alteration in RBL-2H3 cells morphology (Figure 2C), and this concentration was not used for further functional characterization of LiRecTCTP. RBL-2H3 cells degranulation was measured by the beta-hexosaminidase activity assay (Figure 2B), a widely used test, mainly for research purposes. Beta-hexosaminidase is an acid hydrolase that characterizes lysosomal-derived secretory granules that are released during degranulation; the enzyme activity was measured by using p-nitrophenyl *N*-acetyl-beta-D-glucosamine as colorimetric substrate. The degranulation effect is evident when LiRecTCTP is incubated for 2 h with the cells and this activity was dependent of toxin concentration. It is important to highlight that the mutated toxin (LiRecDT1 H12A), produced following the same methodological procedures as LiRecTCTP, was not able to induce beta-hexosaminidase release, ruling out the possibility of contaminants being involved in the effect (Figure 2B). The activity of beta-hexosaminidase after 50 and 100 μg/mL LiRecTCTP treatments was increased two-fold and three-fold, respectively, when compared to the negative control (LiRecDT1 H12A). As shown, 100 μg/mL of LiRecTCTP had a higher degranulation effect than the positive control 48/80, a well-known polymer which triggers mast cell activation. *L. intermedia* crude venom also activates RBL-2H3 cells degranulation in a concentration-dependent manner. It is important to mention the cromolyn inhibitory effect on LiRecTCTP induced beta-hexosaminidase activity. Cromolyn blocks or reduces the mediators released from mast cells, suggesting a pro-inflammatory mechanism of histamine release induced by the LiRecTCTP toxin.

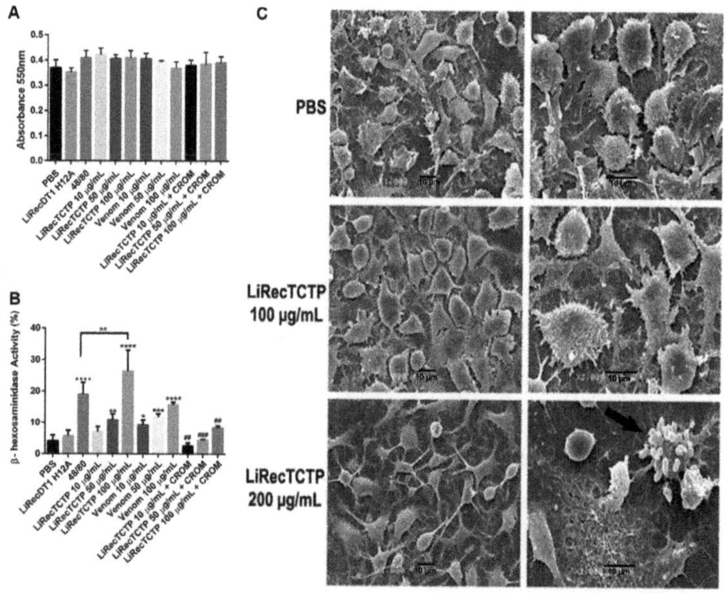

Figure 2. Effects of LiRecTCTP in mast cells viability and degranulation in vitro. RBL-2H3 cells (mast-like cells) were incubated with LiRecTCTP (10, 50, and 100 μg/mL), total venom from *L. intermedia* (10, 50, and 100 μg/mL), compound 48/80 (100 μg/mL) (positive control), LiRecDT1 H12A (100 μg/mL) (negative control) or PBS (negative control). After 2 h viability, morphology and activity of the granular enzyme Beta-hexosaminidase were measured. Inhibition of degranulation was performed using cromolyn (CROM) (20 μM). (**A**) Cell viability was evaluated using MTT assay. The values represent the average of the three independent experiments ± s.d. (performed in pentaplicate). (**B**) Beta-hexosaminidase activity assay. Results are expressed as the percentage of the total beta-hexosaminidase activity present in the cells, after subtracting the activity in the supernatant of unstimulated cells. The values represent the average of the three independent experiments ± s.d. (performed in pentaplicate) (* $p < 0.1$; ** $p < 0.01$; *** $p < 0.001$ and **** $p < 0.0001$ compared with control; ## $p < 0.01$, ### $p < 0.001$ compared to LiRecTCTP treatment). (**C**) Scanning electron microscopy (SEM) of RBL-2H3 cells after 2 h treatment with LiRecTCTP (100 and 200 μg/mL). Images of each sample represent different fields and magnification. Scale bars indicate 10 μm, magnification 1000× (left) and 2000× (right). Arrow: apoptotic cell.

3.3. Effects of LiRecTCTP on the Ca^{2+} Signaling and Cytokines Expression

As changes in the cytosolic Ca^{2+} are central for mast cells activation, we performed an assay to measure the Ca^{2+} influx in RBL-2H3 following a LiRecTCTP treatment (Figure 3A). We could observe a dose-dependent positive effect of LiRecTCTP in the Ca^{2+} influx. Cromolyn abrogated LiRecTCTP effects on Ca^{2+} levels. We also analyzed the cytokine production evoked by LiRecTCTP treatment in RBL-2H3 cells, by measuring the relative mRNA levels for IL-3 (Figure 3B), IL-4 (Figure 3C) and IL-13 (Figure 3D) by means of RT-PCR. LiRecTCTP increased the expression of IL-3, IL-4 and IL-13 in a dose-dependent manner, when compared to the negative controls (PBS and 100 μg/mL of GFP recombinant protein).

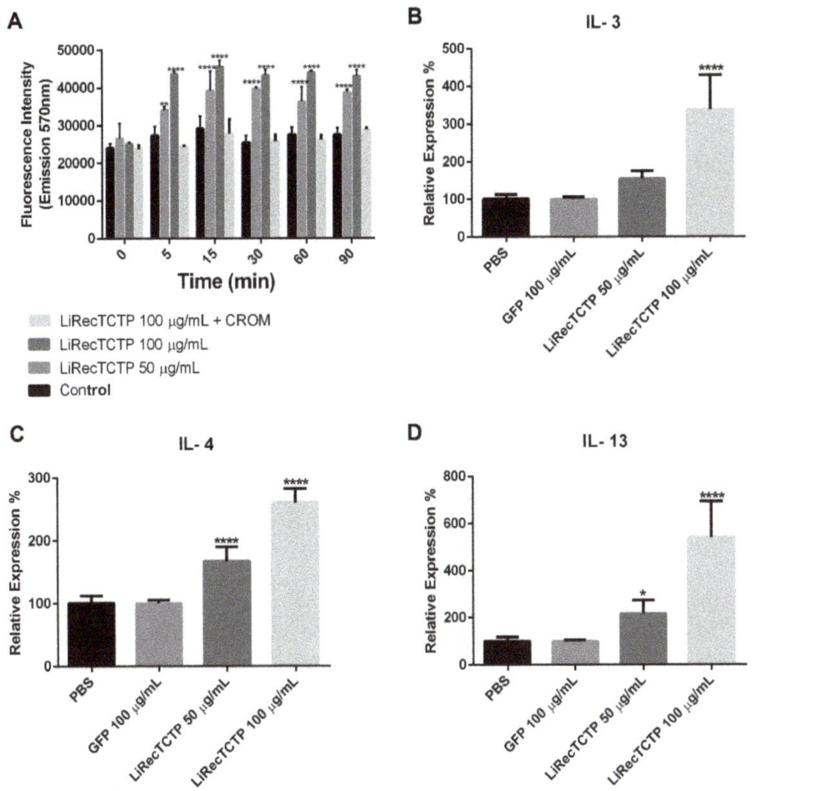

Figure 3. Effect of treatment of RBL-2H3 cells with LiRecTCTP on calcium influx and expression of interleukins. (**A**) RBL-2H3 cells were incubated with LiRecTCTP (50 and 100 μg/mL) in the presence of Fluo-4 AM in buffer containing calcium. The fluorescence of Fluo-4 was measured after 0, 5, 15, 30, 60, and 90 min. As a negative control, cells were incubated without LiRecTCTP. Cromolyn (CROM) (20 μM) inhibitory effect was evaluated in the presence of LiRecTCTP (100 μg/mL). The values represent the average of the three experiments ± s.d. (** $p < 0.01$ and **** $p < 0.0001$). (**B**) Quantitative real-time PCR of IL-3 mRNA levels in RBL-2H3 exposed or not to LiRecTCTP (50 and 100 μg/mL) and GFP (100 μg/mL). (**C**) Quantitative real-time PCR of IL-4 mRNA levels in RBL-2H3 exposed or not to LiRecTCTP (50 and 100 μg/mL). (**D**) Quantitative real-time PCR of IL-13 mRNA levels in RBL-2H3 exposed or not to LiRecTCTP (50 and 100 μg/mL) and GFP (100 μg/mL). For quantification, we used the ΔΔCt method with GAPDH as an endogenous control for each sample (* $p < 0.1$ and **** $p < 0.0001$, compared with controls, PBS and GFP recombinant protein). Data represent mean ± s.d. of three independent experiments.

3.4. LiRecTCTP In Vivo Effects—Vascular Permeability and Edema

In order to evaluate the effect of different histamine receptor blockers on the histaminergic response triggered by LiRecTCTP, we performed two animal studies in which vascular permeability and edema formation were assessed. Well-established antihistaminic drugs with different targets were used in these experiments: promethazine (PRO), an H1 receptor antagonist; cimetidine (CIM), an H2 receptor antagonist; thioperamide (THIO), acts on H3 and H4 receptors; and cromolyn (CROM), a mast cell degranulation blocker. Vascular permeability was measured by Evans blue extravasation (Figure 4A) after intradermic inoculation of LiRecTCTP in mice, previously treated with an antihistaminic or not. Quantification was performed after Evans blue elution (Figure 4B). Images from mice skin and amount of dye eluted show that cromolyn was the most effective drug to reduce LiRecTCTP effects on vascular permeability (absorbance of the eluted dye was very similar to the negative control, PBS). Promethazine and thioperamide could inhibit about 30% of the LiRecTCTP effect in increasing vascular permeability. Cimetidine did not alter the increase in vascular leakage of Evans dye caused by LiRecTCTP.

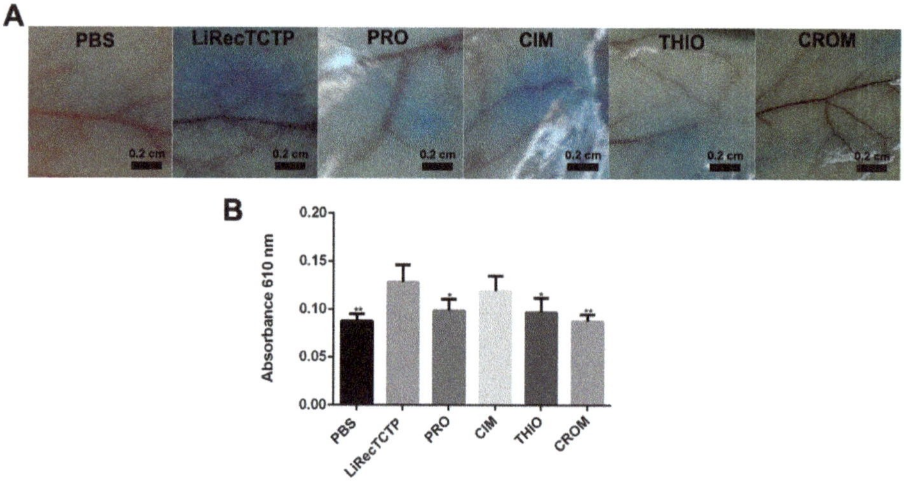

Figure 4. Effect of treatment with the mast cell degranulation inhibitor or histamine receptor antagonists on vascular permeability induced by LiRecTCTP. Mice (n = 5) were pre-treated with promethazine (PRO), cimetidine (CIM), thioperamide (THIO), cromolyn (CROM), or PBS (control). Animals received solution of Evans blue dye in PBS intravenously prior to intradermal injection of LiRecTCTP (10 µg) or PBS (control). (**A**) Representative images of mice dorsal skin at the point of sample inoculation. (**B**) Quantitative measurement of dye leakage induced by LiRecTCTP and inhibition with CROM, PRO, CIM, and THIO. Data represent mean ± s.e.m of one representative experiment from three independent biological replicates (* $p < 0.1$ and ** $p < 0.01$).

We also used mice to evaluate if LiRecTCTP edematogenic effects could be inhibited by the anti-histaminic drugs; the effect of these inhibitors is shown on Figure 5A compared to LiRecTCTP by itself. Cimetidine did not present a significant effect on the paw edema caused by LiRecTCTP; a small inhibition of edema is observed in the first 10 min after toxin administration (Figure 5D). As shown for vascular permeability, promethazine, thioperamide, and cromolyn prevented LiRecTCTP effects on mice paw (Figure 5B,C,E). The time-courses for promethazine and thioperamide had the same profile (Figure 5A). When we compared the effects of these drugs on the paw edema generated by LiRecTCTP, we observed a decrease along the time (5–1440 min). Among tested drugs, cromolyn showed the highest inhibition of LiRecTCTP-induced edema (Figure 5A,E). Figure 5B–E show the anti-histaminic effects and their comparison with respective controls: (i) LiRecTCTP or PBS inoculation in animals previously treated with the drug and (ii) preliminary treatment with PBS and following inoculation of PBS or LiRecTCTP in mice paw. These graphs show there was no significant swelling of the paw after the inoculation of the same volume of PBS and also that the anti-histaminic drugs did not cause unspecific edema. When we compare LiRecTCTP edema curves in the presence of cromolyn and the negative control (PBS), they are very similar (Figure 5E). As observed in permeability assay, cromolyn abrogates LiRecTCTP edematogenic effects.

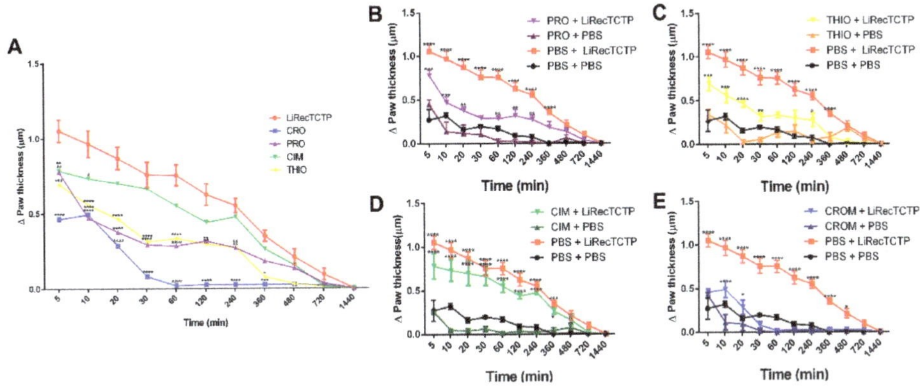

Figure 5. Effect of the treatment with the mast cell degranulation inhibitor or histamine receptor antagonists on edema induced by LiRecTCTP. Mice ($n = 5$) were pre-treated with promethazine (PRO), cimetidine (CIM), thioperamide (THIO), cromolyn (CROM), or PBS (control), and thereafter, animals were injected with LiRecTCTP (10 μg) or PBS (control) into footpads for edema. (**A**) Paw edema observed after injection with LiRecTCTP, in animals previously treated with PBS (LiRecTCTP), mast cell degranulation inhibitor (CRO), or histamine receptor antagonists (PRO, CIM, THIO). (**B**) Paw edema observed after injection with LiRecTCTP or PBS, in animals previously treated with PBS or promethazine (PRO). (**C**) Paw edema observed after injection with LiRecTCTP or PBS, in animals previously treated with PBS or thioperamide (THIO). (**D**) Paw edema observed after injection with LiRecTCTP or PBS, in animals previously treated with PBS or cimetidine (CIM). (**E**) Paw edema observed after injection with LiRecTCTP or PBS, in animals previously treated with PBS or cromolyn (CROM). Values represent the thickness difference between the edema after injection with LiRecTCTP and initial before injections. Each point represents the mean ± s.e.m of five animals from one representative experiment from three independent biological replicates (* $p < 0.1$; ** $p < 0.01$; *** $p < 0.001$ and **** $p < 0.0001$).

3.5. *LiRecTCTP In Vivo Effects—Dermonecrotic Lesion*

We analyzed the role of LiRecTCTP in the dermonecrotic lesion provoked by *Loxosceles* spiders bite accidents. These necrotic lesions are the hallmark of cutaneous manifestation of *Loxosceles* envenomation. The most studied class of *Loxosceles* toxins is phospholipases-D (also called dermonecrotic toxins) whose biological effects can reproduce the ones observed by crude venom inoculation in rabbit skin [7]. In these experiments, we used LiRecTCTP together with an isoform of *L. intermedia* dermonecrotic toxin, LiRecDT1 [28], to evaluate the synergistic action of these toxins. As a negative control, we used an inactive recombinant protein, GFP, which was produced and purified using the same conditions as LiRecTCTP and LiRecDT1. After the subcutaneous inoculation of toxins in rabbit skin, the site was photographed at the time of inoculation and after 24 h for macroscopic evaluation (Figure 6A). After this time, skin patches were collected and processed for microscopic evaluation by histology analyses (Figure 7). Negative controls show that inoculation (PBS or GFP protein) did not trigger any macroscopic (Figure 6A) or microscopic (Figure 7A,A1) alterations in rabbit skin during our experiment. PBS control did not alter normal skin histology (data not show). Dermonecrotic toxin (LiRecDT1) triggered the hallmark lesion, presenting gravitational spreading (Figure 6A) and a marked inflammatory response, which can be observed in the histological analyses by a great number of neutrophils around blood vessels and diffused to connective tissue surrounding the inoculating site, and development of edema (Figure 7B,B1). As expected, LiRecTCTP alone did not cause a skin lesion, but dose-dependent erythema and edema at inoculation site were observed (Figure 6A). The swelling caused by the toxin is visualized in the histology sections, where the width of the skin is greater in LiRecTCTP samples (Figure 7C,D), when compared to GFP recombinant protein (Figure 7A, negative control) and LiRecDT1 (Figure 7B). Image analysis of histological sections revealed that LiRecDT1 promoted an increase of 41% in the tissue area (from epidermis to adipose) when compared to GFP. LiRecTCTP triggered increased edema of 87% and 91% (compared to GFP), when 10 and 20 µg were used, respectively. The combination of LiRecDT1 and LiRecTCTP 20 µg provoked an increased edema area of 340% when compared to GFP.

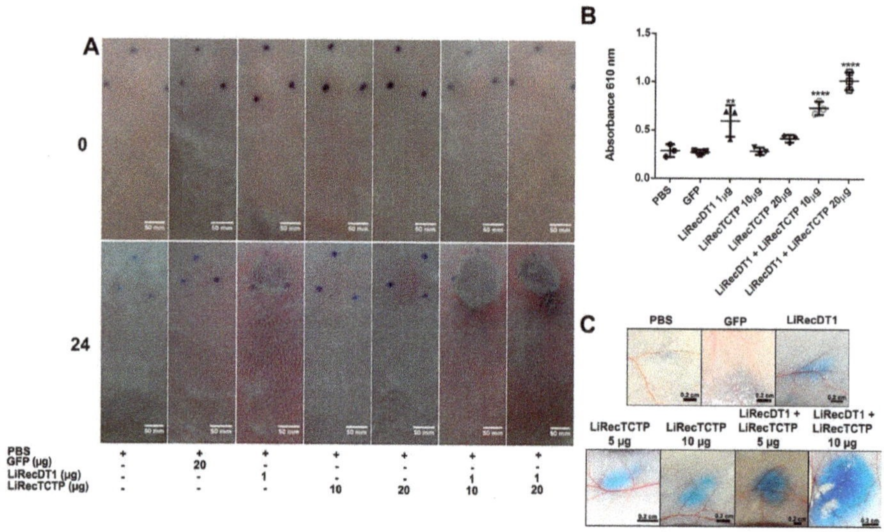

Figure 6. Inflammatory response of combined recombinant toxins (LiRecTCTP and LiRecDT1) in vivo. (**A**) Macroscopic evaluation of rabbit skin exposed to recombinant toxins (LiRecTCTP, LiRecDT1, or combined toxins LiRecTCTP/LiRecDT1). Rabbits were subcutaneously injected with dermonecrotic toxin LiRecDT1 (1 µg, as positive control), LiRecTCTP (10 and 20 µg), LiRecDT1 (1 µg) combined with LiRecTCTP (10 and 20 µg), GFP (20 µg), a recombinant inactive protein (negative control), or PBS (negative control) (+, present; −, absent). Animal skins were photographed just after inoculation (0 h) and 24 h following injection. The same animal received the seven samples for adequate comparison, experiment was repeated twice, using 2 and 4 rabbits respectively. (**B**) Inflammatory reactions induced by toxins and controls were estimated by measurement of myeloperoxidase activity from neutrophils infiltrate at dermis. Values are expressed as mean ± s.e.m of absorbance at 610 nm. Each point represents the average of three replicates from the inoculation site on rabbit skin at the end of experiment (24 h) (** $p < 0.01$ and **** $p < 0.0001$). (**C**) Effect of LiRecTCTP and LiRecDT1 on vascular permeability of skin vessels. Mice were injected intradermally with of LiRecTCTP (5 or 10 µg), LiRecDT1 (1 µg), or recombinant GFP (10 µg) (negative control). PBS was used as a vehicle control. Experiment was performed three times using groups of five mice for each condition. Dye leakage induced by LiRecTCTP combined with LiRecDT1 is higher than the leakage observed with each toxin alone. Scale bar points 0.2 cm.

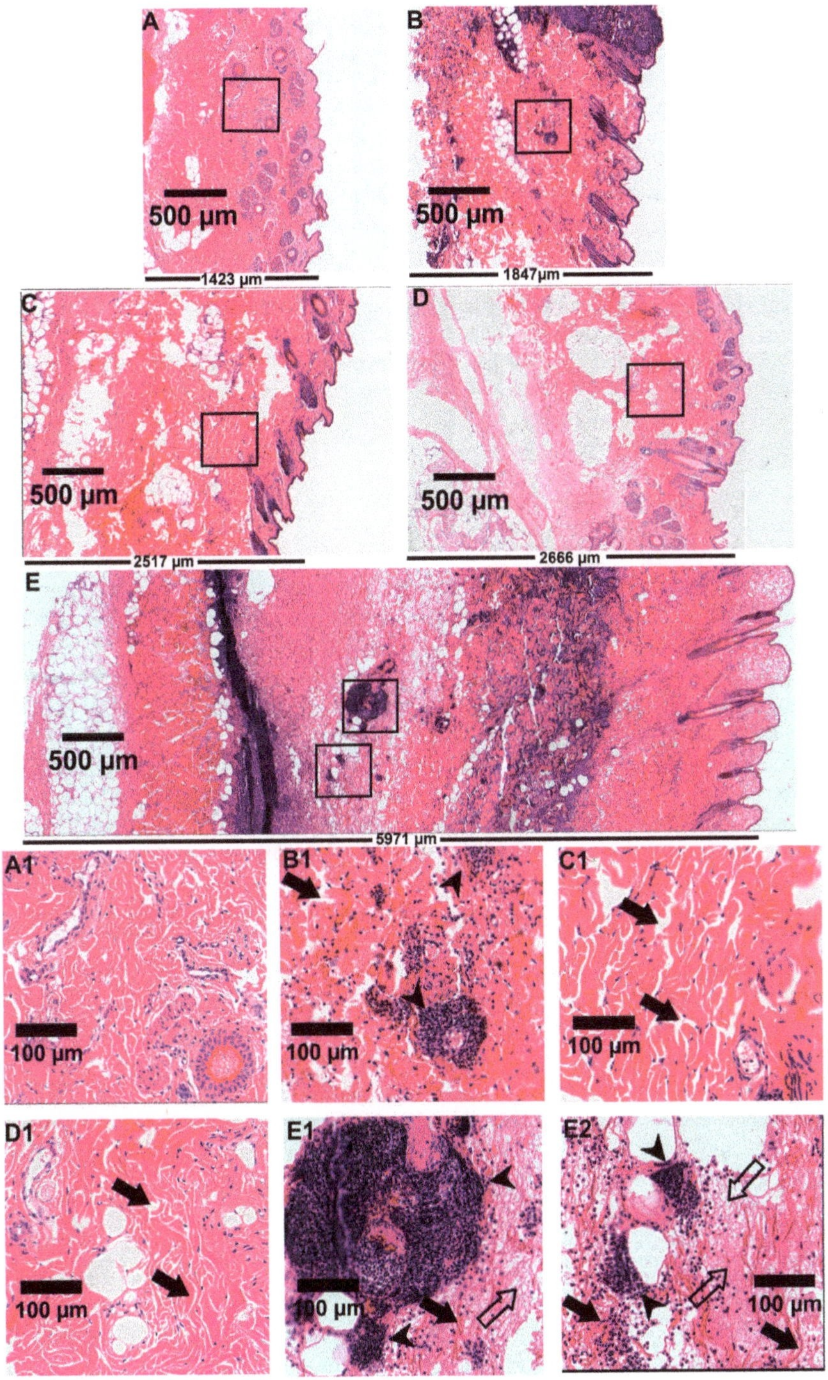
A
500 µm
1423 µm
B
500 µm
1847µm
C
500 µm
2517 µm
D
500 µm
2666 µm
E
500 µm
5971 µm
A1
100 µm
B1
100 µm
C1
100 µm
D1
100 µm
E1
100 µm
E2
100 µm

Figure 7. Microscopic evaluation of rabbit skin exposed to recombinant toxins (LiRecTCTP, LiRecDT1, or combined toxins LiRecTCTP/LiRecDT1). Light microscopic analysis of tissue sections was performed on rabbit skin after 24 h of injection. The tissue sections were stained with hematoxylin and eosin. Edema triggered in rabbit skin by (**A**) GFP, (**B**) LiRecDT1 (1 μg), (**C**) LiRecTCTP (10 μg), (**D**) LiRecTCTP (20 μg), and (**E**) the combination of LiRecDT1 (1 μg) and LiRecTCTP (20 μg), as visualized by skin thickness. Skin structures are compared via scanning of images from epidermal (on the right of figure) to muscular tissues (on the left of figure) under the same laboratory conditions (Scale bars indicate 500 μm). The width of the tissue (**E**) points to a deep edema after LiRecTCTP and LiRecDT1 combination when compared to toxins alone (**B**, **D**). Isolated LiRecTCTP (**C**, **D**) induced a higher edema compared to LiRecDT1 alone (**B**) or negative control GFP (**A**), which shows a normal skin histology (**A1**). An intense inflammatory response with the presence of neutrophils and fibrinoid exudates into the dermis is shown when both toxins were administered (**E1**, **E2**) compared to isolated LiRecTCTP (**C1**, **D1**) or LiRecDT1 (**B1**) (Scale bars indicate 100 μm). Closed arrows indicate disorganization of collagen fibers and dermal edema, closed arrowheads indicate a massive inflammatory response with the presence of neutrophils, and open arrows indicate fibrin network deposition. Thickness of the skin tissue (**A–E**) is shown in the bottom of each tissue section (μm).

Furthermore, dermal edema and disorganization and separation of collagen fibers provide evidence for this swelling (Figure 7C1,D1). The combined use of LiRecDT1 and LiRecTCTP toxins has a synergistic effect on dermonecrosis development—a higher gravitational spreading (Figure 6A) and an increased inflammatory response (Figure 7E) when compared to LiRecDT1 alone (Figure 7B). There are more leucocytes recruited at the lesion site, notably in the papillary dermis, when LiRecTCTP is injected together with the dermonecrotic toxin (Figure 7E), and this number is directly dependent on the dose. Capillary changes with increased permeability, resulting in the passage of plasma into the connective tissue, evidenced by fibrin network formation, is also observed in the presence of both toxins (Figure 7E,E1).

The edema observed in the presence of both toxins was enormous; although macroscopic images do not reveal clearly, this aspect can be verified by respective histology sections, e.g., massive increase in the width of the skin, disorganization of collagen fibers and dermal edema (Figure 7E). In order to quantify the inflammatory response triggered by the toxins in rabbit skin, we evaluate myeloperoxidase activity in the skin patches that were collected (Figure 6B). We can observe an increase of 25% in myeloperoxidase activity when LiRecDT1 was combined with 10 μg LiRecTCTP when compared to LiRecDT1 alone. When 20 μg of LiRecTCTP was used, the increase reaches 65%. LiRecTCTP alone has no significant effect on myeloperoxidase activity; these results were similar to the negative controls (PBS and GFP). We also investigated the synergy of LiRecDT1 and LiRecTCTP in the vascular permeability using mice. Figure 6C shows representative images from Evans Blue assay; dye leakage can be observed after the inoculation of both toxins. This permeability is increased when toxins are administered together, and this was dependent of LiRecTCTP dose. Combination of 10 μg of LiRecTCTP and 1 μg LiRecDT1 administration resulted in a huge and intense blue spot at the inoculation site. There was no relevant extravasation of Evans blue in negative controls (PBS and GFP). A recombinant GFP obtained in the same heterologous system as the *Loxosceles* toxins was used to rule out an unspecific effect due to bacterial contaminants.

4. Discussion

Clinical symptoms of envenomation include histamine-related responses; although they are less frequent, there are reports of hypersensitivity or even allergic reactions after spider bites [9,10]. The presence of histamine in the envenomation site of Loxoscelism can cause edema and endothelial changes such as increased vascular permeability and vasodilation, which contribute to the systemic dispersion of venom components and exacerbate the inflammatory response triggered by the bite [7,8]. Inflammatory responses can be related to both mast cells and histamine [6,11]. It has already been shown that *L. intermedia* venom increases vascular permeability and induces vascular relaxation in rats [29] and that these effects are related to the ability of venom to degranulate mast cells and release mediators such as histamine [11].

Different studies that investigated the protective effects of recombinant *Loxosceles* phospholipases-D (PLDs) or even the neutralizing effects of serum produced with these toxins found that the edematogenic activity of *Loxosceles* venoms is particularly difficult to neutralize and raised the possibility that other venom components may be responsible for edema development [30,31]. LiRecTCTP induces an increase in microvascular permeability of skin vessels and is a component of edema formation with an earlier and faster effect when compared to the inflammatory response triggered by whole *L. intermedia* venom in mouse paws [2]. The effects of *L. intermedia* crude venom on RBL-2H3 degranulation (Figure 2C) are related to the histaminergic effects of Loxoscelism. The report of TCTP-related toxins in the *Loxosceles* venoms were previously done by transcriptomic studies using venom gland transcripts [1], by cloning and recombinant expression of LiRecTCTP [2], and recently, TCTP was identified in the proteomic study of the *L. intermedia* venom [32]. Additionally, TCTP was immunodetected in the whole venom of *Loxosceles* species (*L. intermedia, L. gaucho,* and *L. laeta*) [3], pointing to conservation and biological relevance. There are described TCTPs from other spiders [33,34], although they are not related to venomous accidents as *Loxosceles* [35].

The toxins purification was a critical procedure for this study: pure and correctly folded LiRecTCTP was obtained (Figure 1). Two other recombinant proteins were used as negative controls to exclude an effect due to a possible prokaryotic contamination from the heterologous expression system: green fluorescent protein (GFP), an innocuous protein, and LiRecDT1 H12A, a mutated isoform derived from LiRecDT1, with drastically reduced activity to residual levels [2,27,36].

In this study of LiRecTCTP's biological role, it was relevant to rule out its possible cytotoxic effects on RBL-2H3, which could cause release of cellular contents and mimic degranulation in spite of a mechanism dependent of LiRecTCTP action. The morphology of RBL-2H3 cells was not altered by LiRecTCTP, cells submitted to treatment with 100 µg/mL did not differ from control cells, and they remained adhered and spread onto the slide, showing philopodia (Figure 2C). The unreduced capacity to metabolize MTT by cells treated with LiRecTCTP implies a specific effect on degranulation of RBL-2H3 cells (Figure 2A). Beta-hexosaminidase activity after LiRecTCTP treatment of RBL-2H3 cells infers that this toxin can directly trigger degranulation process in these basophil lineage cells (Figure 2B). Cromolyn is known as a "mast cell stabilizer." It interferes with the release of inflammatory and other chemical mediators from mast cells and either blocks or reduces the amount released [37]. The cromolyn effect as a mast cell stabilizer is dose-dependent.

Human TCTP was already well-characterized as a histaminergic molecule [38–40] and suggested as a putative target for therapeutics in asthma and allergy [41,42]. Several reports indicate TCTP involvement in the inflammatory response of parasite-infected individuals [43–46].

Activation of basophils and mast cells can be monitored through different approaches, including morphological changes, phenotypic changes, and quantification of secreted mediators. IL-4 is produced by basophils in the early response to a stimulus, and IL-3, produced by mast cells, is involved in severe hypersensitivity reactions. Functionally, mast cells, and basophils overlap in their ability to produce several mediators, including histamine and granule proteases. IL-3, IL-4 and IL-13 cytokines act as immunomodulators of other immune cells in the inflammatory and allergic signaling pathway and play pivotal roles in exacerbating the inflammatory responses in vivo [47,48]. Expression of IL-3, IL-4 and IL-13 was induced by LiRecTCTP in RBL-2H3 cells, which indicates that this protein could be involved in the cutaneous inflammatory and histaminic conditions of *Loxosceles* envenomation (Figure 3B). As these cytokines are able to recruit inflammatory immune cells to the bite site, LiTCTP could contribute to the exacerbated inflammatory response by stimulating the production and release of these cytokines. The expression of these cytokines by the RBL-2H3 cells is related to their activation by LiRecTCTP, as indicated by increased Ca^{+2} influx and beta-hexosaminidase activity (Figure 3A,B).

We used different histamine receptors inhibitors in order to evaluate histaminergic effects of LiRecTCTP: H1R to H4R [49]. Alterations in vascular permeability, measured by the leakage of administered Evans blue from vessels (Figure 4), and the edematogenic effect of LiRecTCTP (Figure 5) were performed in mice previously treated or not treated with the anti-histaminic drugs. Cimetidine presented minimum inhibition effect of LiRecTCTP; it could not block toxin-induced permeability and modestly reduced paw edema only in the first 10 min after toxin inoculation. Absence of cimetidine inhibition is explained by the fact that its targets (H2 receptors) are mainly localized in the stomach, brain, and cardiac tissue, but typically not in the skin. Effects of promethazine are related to the blockage of H1R activation by histamine, the previous treatment with this drug resulted in inhibition of LiRecTCTP action on permeability and edema. H1 receptors are expressed by a broad range of cells, including airway and vascular smooth muscle cells, endothelial cells, monocytes, neutrophils, and T and B cells [49]. Thioperamide is a dual H3-H4 receptors antagonist. H3 receptors are irrelevant in the context of our experiments as these histamine receptors are almost exclusively expressed in the nervous system.

On the other hand, histamine H4 receptors are described mainly expressed in cells of the human immune system and influence their cytokine production mediating several effects on chemotaxis [50]. Thioperamide treatment significantly decreased mice response to the histaminergic effects of LiRecTCTP in vascular permeability and paw edema. When the effects of LiRecTCTP on histamine release were blocked by cromolyn, toxin biological effects in animals were almost abrogated: results were similar to the negative controls. These results obtained in animal models point to an in vivo histamine-induced effect by LiRecTCTP, related to mast cell degranulation and histamine effects on H1 and H4 receptors. It is important to mention, in the regard of the pro-inflammatory response triggered by venom and observed in Loxoscelism, that IL-3, IL-4, and histamine can upregulate H1 receptor gene expression. This positive feedback loop could contribute to the exacerbated inflammatory condition seen in dermonecrotic lesions resulted from *Loxosceles* bites. The involvement of H4 receptor in the histaminergic effects of LiRecTCTP should also be highlighted as this receptor is emerging as an essential receptor for the chemoattraction of immunologically relevant cells, contributing to an extended inflammatory cascade.

The investigation of LiRecTCTP participation in dermonecrosis lesion development and clinical features was performed using a well-established in vivo protocol. LiRecDT1 inoculation into rabbit skin, as expected and well-described, caused a characteristic dermonecrosis: a lesion with gravitational spreading, leukocyte infiltration of dermis with the prevalence of neutrophils (PMN), and increased capillary permeability in mice [28,51]. When LiRecTCTP was injected together with the dermonecrotic toxin there was a clear dose-dependent enhancement of all these features at the site of injection: hemorrhage and the gravitational spreading are more intense when compared to LiRecDT1 by itself, and in histopathological analysis, there is an increased number of PMN and more points of fibrin network formation in the connective tissue as a result from an increased microcapillary permeability and disruption of vessel walls (Figure 7). The more prominent cutaneous effect could be explained by the histaminergic effect of LiRecTCTP, which may have contributed to the systemic dispersion of LiRecDT1 toxin. The synergistic effect of both toxins was also revealed by the marked edema and increased dermonecrotic lesions in rabbit skin.

In an envenomation event, LiTCTP is probably acting in edema formation and permeability alterations, corroborating with the huge inflammatory condition of Loxoscelism. LiTCTP could also facilitate the diffusion of other toxins, ultimately promoting venom components spreading from the bite site. Other brown spider toxins that promote extracellular matrix degradation and remodeling, such as metalloproteinases and hyaluronidases, are also implicated in this process [8,52,53]. This combined effect of LiRecDT1 and LiRecTCTP was reproduced in the Evans blue assay. Vascular permeability was highly augmented by the use of both toxins in the inoculation site (Figure 6C). The evaluation of myeloperoxidase activity was crucial to investigate the inflammatory response in mice, as these animals do not develop dermonecrotic lesions (reason still not fully understood) [28,54]. The level of myeloperoxidase (MPO) activity in a sample is directly proportional to the number of neutrophils present, representing the inflammation status [27,55,56]. Results of MPO activity show LiRecTCTP increases the inflammation triggered by LiRecDT1, data that corroborate with its relevant participation in the pro-inflammatory response seen in dermonecrotic lesions (Figure 6B). Concerning the ratio LiRecDT1/LiRecTCTP used in our experiments, transcriptome analysis of *Loxosceles* venoms glands showed that phospholipases-D transcripts such as LiRecDT1 are much more abundant that TCTP transcripts [1]. However, higher concentrations of LiRecDT1 would impair the detection of the synergistic activity between these toxins, inflammatory effects generated only by PLDs would mask the allergenic and inflammatory activities generated by LiTCTP.

Data presented herein confirm previous data that suggested LiRecTCTP histaminergic action by its effect on vascular permeability and edema formation [2]. We suggest that the participation of LiTCTP in the exacerbated inflammatory response is based on LiRecTCTP's direct effect on mast cells and histamine release. Moreover, LiRecTCTP could be considered for potential applications as biotool, as for instance in basophil activation tests or allergy diagnosis and in vitro testing [8,57].

Altogether, the effects observed for LiRecTCTP, resulting in increased inflammatory response, capillary permeability and edema, as well as acting synergistically with dermonecrotic toxin LiRecDT1, unveil LiTCTP's role as an additional spreading factor present in the venom of *Loxosceles* spiders. Along with the classic spreading agents described in the venom [13,53], LiTCTP through its histaminergic mechanisms can contribute to the spread of other toxins from the bite site, accentuating the local and even systemic post-envenoming condition.

Author Contributions: Conceptualization, A.S.-R., S.S.V., O.M.C., and M.B.-F.; Methodology, M.B.-F., K.G.M., A.B.C.B., A.C.M.W., L.P.d.S., L.D., L.V. and B.S.; Analysis, M.B.-F., K.G.M., B.S., K.C.B., A.B.C.B., A.C.M.W., L.P.d.S., L.D., O.M.C., and L.V.; Writing (Original Draft Preparation), A.S.-R. and M.B.-F.; Writing (Review and Editing), A.S.-R., M.B.-F., L.H.G., S.S.V., and O.M.C.; Visualization, M.B.-F.; Supervision, A.S.-R., L.H.G., and S.S.V.; Project Administration, A.S.-R., O.M.C., and L.H.G.; Funding Acquisition, A.S.-R., O.M.C., L.H.G., and S.S.V.

Funding: This work was supported by grants from CAPES, CNPq, UFPR (Federal University of Paraná), FUNDAÇÃO ARAUCÁRIA-PR/SETI-PR/SESA-PR/MS-Decit/PPSUS, Brazil.

Conflicts of Interest: The authors declare no conflict of interest.

References

1. Gremski, L.H.; Da Silveira, R.B.; Chaim, O.M.; Probst, C.M.A.; Ferrer, V.P.; Nowatzki, J.; Weinschutz, H.C.; Madeira, H.M.I.; Gremski, W.; Nader, H.B.; et al. A novel expression profile of the Loxosceles intermedia spider venomous gland revealed by transcriptome analysis. *Mol. Biosyst.* **2010**, *6*, 2403–2416. [CrossRef]
2. Sade, Y.B.; Bóia-Ferreira, M.; Gremski, L.H.; Da Silveira, R.B.; Gremski, W.; Senff-Ribeiro, A.; Chaim, O.M.; Veiga, S.S. Molecular cloning, heterologous expression and functional characterization of a novel translationally-controlled tumor protein (TCTP) family member from Loxosceles intermedia (brown spider) venom. *Int. J. Biochem. Cell Biol.* **2012**, *44*, 170–177. [CrossRef]
3. Buch, D.R.; Souza, F.N.; Meissner, G.O.; Morgon, A.M.; Gremski, L.H.; Ferrer, V.P.; Trevisan-Silva, D.; Matsubara, F.H.; Boia-Ferreira, M.; Sade, Y.B.; et al. Brown spider (Loxosceles genus) venom toxins: Evaluation of biological conservation by immune cross-reactivity. *Toxicon* **2015**, *108*, 154–166. [CrossRef]
4. MacDonald, S.M.; Rafnar, T.; Langdon, J.; Lichtenstein, L.M. Molecular identification of an IgE-dependent histamine-releasing factor. *Science* **1995**, *269*, 688–690. [CrossRef] [PubMed]
5. Kashiwakura, J.I.; Ando, T.; Matsumoto, K.; Kimura, M.; Kitaura, J.; Matho, M.H.; Zajonc, D.M.; Ozeki, T.; Ra, C.; MacDonald, S.M.; et al. Histamine-releasing factor has a proinflammatory role in mouse models of asthma and allergy. *J. Clin. Invest.* **2012**, *122*, 218–228. [CrossRef] [PubMed]
6. Theoharides, T.C.; Alysandratos, K.-D.; Angelidou, A.; Delivanis, D.-A.; Sismanopoulos, N.; Zhang, B.; Asadi, S.; Vasiadi, M.; Weng, Z.; Miniati, A.; et al. Mast cells and inflammation. *Biochim. Biophys. Acta Mol. Basis Dis.* **2012**, *1822*, 21–33. [CrossRef]
7. Gremski, L.H.; Trevisan-Silva, D.; Ferrer, V.P.; Matsubara, F.H.; Meissner, G.O.; Wille, A.C.M.; Vuitika, L.; Dias-Lopes, C.; Ullah, A.; De Moraes, F.R.; et al. Recent advances in the understanding of brown spider venoms: From the biology of spiders to the molecular mechanisms of toxins. *Toxicon* **2014**, *83*, 91–120. [CrossRef] [PubMed]

8. Chaves-Moreira, D.; Matsubara, F.H.; Schemczssen-Graeff, Z.; De Bona, E.; Heidemann, V.R.; Guerra-Duarte, C.; Gremski, L.H.; Chávez-Olórtegui, C.; Senff-Ribeiro, A.; Chaim, O.M.; et al. Brown Spider (Loxosceles) Venom Toxins as Potential Biotools for the Development of Novel Therapeutics. *Toxins (Basel).* **2019**, *11*, 355. [CrossRef] [PubMed]
9. Makris, M.; Spanoudaki, N.; Giannoula, F.; Chliva, C.; Antoniadou, A.; Kalogeromitros, D. Acute generalized exanthematous pustulosis (AGEP) triggered by a spider bite. *Allergol. Int.* **2009**, *58*, 301–303. [CrossRef]
10. Lane, L.; McCoppin, H.H.; Dyer, J. Acute generalized exanthematous pustulosis and coombs-positive hemolytic anemia in a child following Loxosceles reclusa envenomation. *Pediatr. Dermatol.* **2011**, *28*, 685–688. [CrossRef]
11. Paludo, K.S.; Biscaia, S.M.P.; Chaim, O.M.; Otuki, M.F.; Naliwaiko, K.; Dombrowski, P.A.; Franco, C.R.C.; Veiga, S.S. Inflammatory events induced by brown spider venom and its recombinant dermonecrotic toxin: A pharmacological investigation. *Comp. Biochem. Physiol. C Toxicol. Pharmacol.* **2009**, *149*, 323–333. [CrossRef] [PubMed]
12. MacDonald, S.M. Potential role of histamine releasing factor (HRF) as a therapeutic target for treating asthma and allergy. *J. Asthma Allergy* **2012**, *5*, 51–59. [CrossRef] [PubMed]
13. Feitosa, L.; Gremski, W.; Veiga, S.S.; Elias, M.C.Q.B.; Graner, E.; Mangili, O.C.; Brentani, R.R. Detection and characterization of metalloproteinases with gelatinolytic, fibronectinolytic and fibrinogenolytic activities in Brown spider (Loxosceles intermedia) venom. *Toxicon* **1998**, *36*, 1039–1051. [CrossRef]
14. Bradford, M.M. A Rapid and Sensitive Method for the Quantitation of Microgram Quantities of Protein Utilizing the Principle of Protein-Dye Binding. *Anal. Biochem.* **1976**, *72*, 248–254. [CrossRef]
15. Vuitika, L.; Chaves-Moreira, D.; Caruso, I.; Lima, M.A.; Matsubara, F.H.; Murakami, M.T.; Takahashi, H.K.; Toledo, M.S.; Coronado, M.A.; Nader, H.B.; et al. Active site mapping of Loxosceles phospholipases D: Biochemical and biological features. *BBA Mol. Cell Biol. Lipids* **2016**, *1861*, 970–979. [CrossRef]
16. Falcone, F.H.; Wan, D.; Barwary, N.; Sagi-Eisenberg, R. RBL cells as models for in vitro studies of mast cells and basophils. *Immunol. Rev.* **2018**, *282*, 47–57. [CrossRef]
17. Hirano, T.; Koyanagi, A.; Kotoshiba, K.; Shinkai, Y.; Kasai, M.; Ando, T.; Kaitani, A.; Okumura, K.; Kitaura, J. The Fab fragment of anti-IgE Cε2 domain prevents allergic reactions through interacting with IgE-FcεRIα complex on rat mast cells. *Sci. Rep.* **2018**, *8*, 8–15. [CrossRef]
18. Karnovsky, M.J. A formaldehyde-glutaraldehyde fixative of high osmolality for use in electron-microscopy. *J. Cell Biol.* **1965**, *27*, 137–138.
19. Wille, A.C.M.; Chaves-Moreira, D.; Trevisan-Silva, D.; Magnoni, M.G.; Boia-Ferreira, M.; Gremski, L.H.; Gremski, W.; Chaim, O.M.; Senff-Ribeiro, A.; Veiga, S.S. Modulation of membrane phospholipids, the cytosolic calcium influx and cell proliferation following treatment of B16-F10 cells with recombinant phospholipase-D from Loxosceles intermedia (brown spider) venom. *Toxicon* **2013**, *67*, 17–30. [CrossRef]
20. Chaves-Moreira, D.; Souza, F.N.; Fogaça, R.T.H.; Mangili, O.C.; Gremski, W.; Senff-Ribeiro, A.; Chaim, O.M.; Veiga, S.S. The relationship between calcium and the metabolism of plasma membrane phospholipids in hemolysis induced by brown spider venom phospholipase-D toxin. *J. Cell. Biochem.* **2011**, *112*, 2529–2540. [CrossRef]
21. Spandidos, A.; Wang, X.; Wang, H.; Seed, B. PrimerBank: A resource of human and mouse PCR primer pairs for gene expression detection and quantification. *Nucleic Acids Res.* **2009**, *38*, 792–799. [CrossRef] [PubMed]
22. Schmittgen, T.D.; Livak, K.J. Analyzing real-time PCR data by the comparative CT method. *Nat. Protoc.* **2008**, *3*, 1101–1108. [CrossRef] [PubMed]
23. Nascimento, E.B.; Costa, K.A.; Bertollo, C.M.; Oliveira, A.C.P.; Rocha, L.T.S.; Souza, A.L.S.; Glória, M.B.A.; Moraes-Santos, T.; Coelho, M.M. Pharmacological investigation of the nociceptive response and edema induced by venom of the scorpion Tityus serrulatus. *Toxicon* **2005**, *45*, 585–593. [CrossRef] [PubMed]
24. Appel, M.H.; da Silveira, R.B.; Chaim, O.M.; Paludo, K.S.; Silva, D.T.; Chaves, D.M.; da Silva, P.H.; Mangili, O.C.; Senff-Ribeiro, A.; Gremski, W.; et al. Identification, cloning and functional characterization of a novel dermonecrotic toxin (phospholipase D) from brown spider (Loxosceles intermedia) venom. *Biochim. Biophys. Acta Gen. Subj.* **2008**, *1780*, 167–178. [CrossRef] [PubMed]
25. Radu, M.; Chernoff, J. An in vivo assay to test blood vessel permeability. *J. Vis. Exp.* **2013**, 2–5. [CrossRef]

26. Ribeiro, R.O.S.; Chaim, O.M.; da Silveira, R.B.; Gremski, L.H.; Sade, Y.B.; Paludo, K.S.; Senff-Ribeiro, A.; de Moura, J.; Chávez-Olórtegui, C.; Gremski, W.; et al. Biological and structural comparison of recombinant phospholipase D toxins from Loxosceles intermedia (brown spider) venom. *Toxicon* **2007**, *50*, 1162–1174. [CrossRef]
27. Chaim, O.M.; Da Silveira, R.B.; Trevisan-Silva, D.; Ferrer, V.P.; Sade, Y.B.; Bóia-Ferreira, M.; Gremski, L.H.; Gremski, W.; Senff-Ribeiro, A.; Takahashi, H.K.; et al. Phospholipase-D activity and inflammatory response induced by brown spider dermonecrotic toxin: Endothelial cell membrane phospholipids as targets for toxicity. *BBA Mol. Cell Biol. Lipids* **2011**, *1811*, 84–96. [CrossRef]
28. Chaim, O.M.; Sade, Y.B.; Da Silveira, R.B.; Toma, L.; Kalapothakis, E.; Chávez-Olórtegui, C.; Mangili, O.C.; Gremski, W.; Von Dietrich, C.P.; Nader, H.B.; et al. Brown spider dermonecrotic toxin directly induces nephrotoxicity. *Toxicol. Appl. Pharmacol.* **2006**, *211*, 64–77. [CrossRef]
29. Rattmann, Y.D.; Pereira, C.R.; Cury, Y.; Gremski, W.; Marques, M.C.A.; da Silva-Santos, J.E. Vascular permeability and vasodilation induced by the Loxosceles intermedia venom in rats: Involvement of mast cell degranulation, histamine and 5-HT receptors. *Toxicon* **2008**, *51*, 363–372. [CrossRef]
30. Mendes, T.M.; Oliveira, D.; Figueiredo, L.F.M.; Machado-de-Avila, R.A.; Duarte, C.G.; Dias-Lopes, C.; Guimarães, G.; Felicori, L.; Minozzo, J.C.; Chávez-Olortegui, C. Generation and characterization of a recombinant chimeric protein (rCpLi) consisting of B-cell epitopes of a dermonecrotic protein from Loxosceles intermedia spider venom. *Vaccine* **2013**, *31*, 2749–2755. [CrossRef]
31. Duarte, C.G.; Bonilla, C.; Guimarães, G.; Machado De Avila, R.A.; Mendes, T.M.; Silva, W.; Tintaya, B.; Yarleque, A.; Chávez-Olórtegui, C. Anti-loxoscelic horse serum produced against a recombinant dermonecrotic protein of Brazilian Loxosceles intermedia spider neutralize lethal effects of Loxosceles laeta venom from Peru. *Toxicon* **2015**, *93*, 37–40. [CrossRef] [PubMed]
32. Trevisan-Silva, D.; Bednaski, A.V.; Fischer, J.S.G.; Veiga, S.S.; Bandeira, N.; Guthals, A.; Marchini, F.K.; Leprevost, F.V.; Barbosa, V.C.; Senff-Ribeiro, A.; et al. A multi-protease, multi-dissociation, bottom-up-to-top-down proteomic view of the Loxosceles intermedia venom. *Sci. Data* **2017**, *4*, 1–7. [CrossRef] [PubMed]
33. Kimura, T.; Ono, S.; Kubo, T. Molecular cloning and sequence analysis of the cDNAS encoding toxin-like peptides from the venom glands of tarantula Grammostola rosea. *Int. J. Pept.* **2012**, *2012*. [CrossRef] [PubMed]
34. Zobel-Thropp, P.A.; Correa, S.M.; Garb, J.E.; Binford, G.J. Spit and venom from scytodes spiders: A diverse and distinct cocktail. *J. Proteome Res.* **2014**, *13*, 817–835. [CrossRef] [PubMed]
35. Senff-Ribeiro, A. Translattionaly controlled tumor protein (TCTP/HRF) in Animal Venoms. In *TCTP/tpt1-Remodeling Signaling from Stem Cell to Disease. Results and Problems in Cell Differentiation.*; Telerman, A., Amson, R., Eds.; Springer International Publishing: Basel, Switzerland, 2017; pp. 193–200.
36. Kusma, J.; Chaim, O.M.; Wille, A.C.M.; Ferrer, V.P.; Sade, Y.B.; Donatti, L.; Gremski, W.; Mangili, O.C.; Veiga, S.S. Nephrotoxicity caused by brown spider venom phospholipase-D (dermonecrotic toxin) depends on catalytic activity. *Biochimie* **2008**, *90*, 1722–1736. [CrossRef] [PubMed]
37. Brogden, R.N.; Speight, T.N.; Avery, G.S. Sodium Cromoglycate (Cromolyn Sodium): A Review of its Mode of Action, Pharmacology, Therapeutic Efficacy and Use. *Drugs* **1974**, *7*, 164–282. [CrossRef]
38. Kawakami, T.; Kashiwakura, J.I.; Kawakami, Y. Histamine-releasing factor and immunoglobulins in asthma and allergy. *Allergy, Asthma Immunol. Res.* **2014**, *6*, 6–12. [CrossRef]
39. Bheekha-Escura, R.; MacGlashan, D.W.; Langdon, J.M.; MacDonald, S.M. Human recombinant histamine-releasing factor activates human eosinophils and the eosinophilic cell line, AML14-3D10. *Blood* **2000**, *96*, 2191–2198. [CrossRef]
40. Kawakami, T.; Blank, U. From IgE to Omalizumab. *J. Immunol.* **2016**, *197*, 4187–4192. [CrossRef]
41. Bommer, U.A.; Thiele, B.J. The translationally controlled tumour protein (TCTP). *Int. J. Biochem. Cell Biol.* **2004**, *36*, 379–385. [CrossRef]
42. Kim, M.; Maeng, J.; Lee, K. Dimerization of TCTP and its clinical implications for allergy. *Biochimie* **2013**, *95*, 659–666. [CrossRef]
43. Gnanasekar, M.; Rao, K.V.N.; Chen, L.; Narayanan, R.B.; Geetha, M.; Scott, A.L.; Ramaswamy, K.; Kaliraj, P. Molecular characterization of a calcium binding translationally controlled tumor protein homologue from the filarial parasites Brugia malayi and Wuchereria bancrofti. *Mol. Biochem. Parasitol.* **2002**, *121*, 107–118. [CrossRef]

44. Rao, K.V.N.; Chen, L.; Gnanasekar, M.; Ramaswamy, K. Cloning and characterization of a calcium-binding, histamine-releasing protein from Schistosoma mansoni. *J. Biol. Chem.* **2002**, *277*, 31207–31213. [CrossRef] [PubMed]
45. Eichhorn, T.; Winter, D.; Büchele, B.; Dirdjaja, N.; Frank, M.; Lehmann, W.D.; Mertens, R.; Krauth-Siegel, R.L.; Simmet, T.; Granzin, J.; et al. Molecular interaction of artemisinin with translationally controlled tumor protein (TCTP) of Plasmodium falciparum. *Biochem. Pharmacol.* **2013**, *85*, 38–45. [CrossRef] [PubMed]
46. MacDonald, S.M.; Bhisutthibhan, J.; Shapiro, T.A.; Rogerson, S.J.; Taylor, T.E.; Tembo, M.; Langdon, J.M.; Meshnick, S.R. Immunerfstimimicry in malaria: Plasmodium falciparum secretes a functional histamine-releasing factor homolog in vitro and in vivo. *Proc. Natl. Acad. Sci. USA.* **2001**, *98*, 10829–10832. [CrossRef]
47. Yamanishi, Y.; Karasuyama, H. Basophil-derived IL-4 plays versatile roles in immunity. *Semin. Immunopathol.* **2016**, *38*, 615–622. [CrossRef]
48. Chhiba, K.D.; Hsu, C.-L.; Berdnikovs, S.; Bryce, P.J. Transcriptional Heterogeneity of Mast Cells and Basophils upon Activation. *J. Immunol.* **2017**, *198*, 4868–4878. [CrossRef]
49. O'Mahony, L.; Akdis, M.; Akdis, C.A. Regulation of the immune response and inflammation by histamine and histamine receptors. *J. Allergy Clin. Immunol.* **2011**, *128*, 1153–1162. [CrossRef]
50. Walter, M.; Kottke, T.; Stark, H. The histamine H4 receptor: Targeting inflammatory disorders. *Eur. J. Pharmacol.* **2011**, *668*, 1–5. [CrossRef]
51. Bertoni da Silveira, R.; Pigozzo, R.B.; Chaim, O.M.; Appel, M.H.; Dreyfuss, J.L.; Toma, L.; Mangili, O.C.; Gremski, W.; Dietrich, C.P.; Nader, H.B.; et al. Molecular cloning and functional characterization of two isoforms of dermonecrotic toxin from Loxosceles intermedia (Brown spider) venom gland. *Biochimie* **2006**, *88*, 1241–1253. [CrossRef]
52. Trevisan-Silva, D.; Bednaski, A.V.; Gremski, L.H.; Chaim, O.M.; Veiga, S.S.; Senff-Ribeiro, A. Differential metalloprotease content and activity of three Loxosceles spider venoms revealed using two-dimensional electrophoresis approaches. *Toxicon* **2013**, *76*, 11–22. [CrossRef] [PubMed]
53. Ferrer, V.P.; de Mari, T.L.; Gremski, L.H.; Trevisan Silva, D.; da Silveira, R.B.; Gremski, W.; Chaim, O.M.; Senff-Ribeiro, A.; Nader, H.B.; Veiga, S.S. A Novel Hyaluronidase from Brown Spider (Loxosceles intermedia) Venom (Dietrich's Hyaluronidase): From Cloning to Functional Characterization. *PLoS Negl. Trop. Dis.* **2013**, *7*, 1–12. [CrossRef] [PubMed]
54. Domingos, M.O.; Barbaro, K.C.; Tynan, W.; Penny, J.; Lewis, D.J.M.; New, R.R.C. Influence of sphingomyelin and TNF-α release on lethality and local inflammatory reaction induced by Loxosceles gaucho spider venom in mice. *Toxicon* **2003**, *42*, 471–479. [CrossRef]
55. Bradley, P.P.; Priebat, D.A.; Christensen, R.D.; Rothstein, G. Measurement of cutaneous inflammation: Estimation of neutrophil content with an enzyme marker. *J. Invest. Dermatol.* **1982**, *78*, 206–209. [CrossRef]
56. Gomez, H.F.; Greenfield, D.M.; Miller, M.J.; Warren, J.S. Direct Correlation between Diffusion of Loxosceles recluse venom and extent of dermal inflammation. *Acad. Emerg. Med.* **2001**, *8*, 309–314. [CrossRef]
57. Konstantinou, G.N.; Asero, R.; Ferrer, M.; Knol, E.F.; Maurer, M.; Raap, U.; Schmid-Grendelmeier, P.; Skol, P.S.; Grattan, C.E.H. EAACI taskforce position paper: Evidence for autoimmune urticaria and proposal for defining diagnostic criteria. *Allergy Eur. J. Allergy Clin. Immunol.* **2013**, *68*, 27–36. [CrossRef]

cells

MDPI

Editorial

Obituary for Susan M. MacDonald, M.D.

Alkis Togias [1,*], Marshall Plaut [1], Jackie Langdon [2], Ulrich-Axel Bommer [3] and Adam Telerman [4,*]

1 Allergy, Asthma and Airway Biology Branch, Division of Allergy, Immunology and Transplantation, National Institute of Allergy and Infectious Diseases, National Institutes of Health, Bethesda, MD 20892, USA; mplaut@niaid.nih.gov
2 School of Medicine, Division of Geriatric Medicine and Gerontology, Johns Hopkins University, Baltimore, MD 21224, USA; jlangdon@jhmi.edu
3 School of Medicine, Faculty of Science, Medicine & Health, University of Wollongong, Wollongong 2522, Australia; ubommer@uow.edu.au
4 Institut Gustave Roussy, Unite Inserm U981, 94805 Villejuif, France
* Correspondence: togiasa@niaid.nih.gov (A.T.); atelerman@gmail.com (A.T.)

check for updates

Citation: Togias, A.; Plaut, M.; Langdon, J.; Bommer, U.-A.; Telerman, A. Obituary for Susan M. MacDonald, M.D.. *Cells* **2021**, *10*, 87. https://doi.org/10.3390/cells10010087

Received: 24 December 2020
Accepted: 5 January 2021
Published: 7 January 2021

Publisher's Note: MDPI stays neutral with regard to jurisdictional claims in published maps and institutional affiliations.

On 9 September 2020, the Allergy and Immunology community lost a prominent member, Susan MacDonald, after a lengthy illness.

After graduating from Regis College and before attending medical school at the University of Massachusetts, Susan worked as a Research Assistant at the laboratory of K. Frank Austen at Harvard Medical School, where she was introduced to immunology. Following medical school, she trained in Internal Medicine at Johns Hopkins Hospital, where she also served as Assistant Chief of Service (Chief Resident) in 1984–1985. Her subspecialty clinical training was in rheumatology, but her research training took place at the Laboratory of Larry Lichtenstein, in the Division of Allergy and Clinical Immunology, Department of Medicine, both at the Johns Hopkins School of Medicine. She went on to join the faculty at Johns Hopkins and attained the rank of Professor in 2004. In 2013, Dr. MacDonald was appointed interim Chief of the Division of Allergy and Clinical Immunology and held this position until her retirement in 2016.

Dr. Susan MacDonald made breakthrough contributions to the understanding of allergic inflammation. She became a world-recognized expert in basophil and mast cell biology, but her major research achievement was the identification and characterization of the biological activity of a key molecule, histamine releasing factor (HRF). In allergic individuals, exposure to allergen immediately triggers a set of symptoms associated with the release of mast cell-derived chemical mediators such as histamine. However, many people also develop a late phase reaction (LPR), occurring anywhere from 2 to 24 h after the initial reaction. This is characterized by a recurrence of symptoms and the release of histamine and other mediators. LPR is induced experimentally when allergic individuals are exposed to the relevant allergen, in the nose, lung and skin. In the natural setting LPR is probably the mechanism for the persistence of symptoms for hours after exposure to a source of allergen that many patients describe. In cases of systemic anaphylaxis LPR may explain the biphasic symptomatology that a substantial number of patients develop. Even though the allergen is no longer present in LPR, histamine and other mediators are still released. Thus, another molecule, an HRF, is likely to be responsible for triggering the secretion of the LPR-associated mediators by basophils and/or mast cells. The source and clinical relevance of such mediators have been studied for approximately 40 years. Although several HRFs have been identified, work in the mid-1980s suggested that an IgE-dependent HRF, which interacts with IgE on basophils and mast cells, was a key mediator of LPR [1]. Dr. MacDonald developed a strong research program to isolate, identify and clone this IgE-dependent HRF, and in 1995, she succeeded [2]. HRF functions by interacting with IgE molecules, but HRF works on only certain subsets of IgE molecules (called IgE+) and not on other subsets. It is possible that HRF binds to specific variable regions of IgE and therefore does not bind to all IgE molecules. This may explain why LPR is not a phenomenon that every patient with allergies develops. The importance of HRF in allergic inflammation is not fully understood, but more recent research, especially by Dr. Toshiaki Kawakami and his colleagues, indicates that HRF is critical for inducing allergic inflammation, and that agents that block HRF will block animal models of allergic inflammation [3].

The action of HRF on IgE is an extracellular one, but in her original paper [2], Dr MacDonald also showed that HRF is identical to an intracellular protein, called translationally controlled tumor protein (TCTP); alas P21, P23, or later fortilin. TCTP was originally discovered in the early 1980s as a highly regulated protein present in proliferating murine cell lines. However, the functional importance of TCTP remained a mystery for a long time, due to the fact that it does not share sequence homology with any other protein family. Hence, Susan's discovery was of dual importance in the history of the TCTP/HRF proteins—it marks the first description of a functional role of TCTP, and it opened the field for a large number of studies, which investigated the extracellular 'branch' of the protein's function. Meanwhile, also a whole array of intracellular functions has been established, such as cell cycle progression, proliferation, cell survival, as well as its role in the promotion of diseases such as cancer. Most of these functional activities have been reviewed in the first 'TCTP book' in 2017, to which Susan MacDonald contributed the chapter 'History of Histamine-Releasing Factor (HRF)/Translationally Controlled Tumor Protein (TCTP) Including a Potential Therapeutic Target in Asthma and Allergy' [4]. In that chapter, Susan clearly expresses her fascination with all aspects of TCTP/HRF, including both its intracellular and extracellular functions. You could see this fascination in the brightness of her eyes, like a kid discovering a new world, inviting the scientists, sharing her work and always ready to help. She dedicated all her life to medicine, science and Johns Hopkins. This kind of motivation and enthusiasm has had a lasting impact on her peers. Her colleague, Professor Judith Karp, at the Johns Hopkins Sidney Kimmel Comprehensive Cancer Center, said about Susan: 'A very talented lady who left an important mark on medicine'.

Known simply as Susan to all who worked with her, she was the first person to turn to when you needed medical advice for a friend or family member. Despite not treating patients for years, her clinical skills were sharp. Dr. Susan MacDonald was a talented, compassionate clinician, and she continued to share these talents with everyone in her circle that needed them.

In 1997, Susan was appointed Deputy Director for Faculty and Career Development at the Johns Hopkins University School of Medicine, Department of Medicine and, in 2002, Associate Chair of Medicine, a position she held for the next 12 years. She was the first woman who held this position at Johns Hopkins. A major responsibility was to oversee the development of more than 575 faculty members and to mentor them through their promotion processes. In this context, she closely mentored more than 35 faculty. Many individuals who Susan mentored remain active and productive members of the scientific community throughout Maryland, and the world. Susan initiated a book that she called "How to Get Promoted at Hopkins" and, in collaboration with the Vice Dean for Faculty Affairs, launched the first edition of the "Silver Book", which helps the Johns Hopkins faculty in their quest for promotion. Her effort and commitment were further reflected in her appointments at the Johns Hopkins School of Medicine Office of Faculty Development, as an Advisor to the School of Medicine Office of Women in Science and in the School of Medicine Advisory Committee on Mentoring. She was recognized for her extraordinary contributions with several awards, including the Department of Medicine David Levine Excellence in Mentoring Award (2003), the first Vice Dean's Leadership Award for the Advancement of Women at Johns Hopkins University School of Medicine (2009). From the American Academy of Allergy, Asthma and Immunology (AAAAI), she received the Women Physicians Leadership in Allergy Award (2001), Women's Involvement Special Recognition Award (2005) and the Gail G. Shapiro Honorary Special Recognition Award (2008). Between 2012 and 2014, she chaired the Leadership Institute and, in 2014–2015, the Task Force on Retention, Inclusion and Diversity of the AAAAI.

Susan MacDonald was also our friend. We spent wonderful moments together and with her beloved husband, Dr. David Herron, and we experienced her warm support in difficult moments. She will be fondly remembered.

Disclaimer: Drs. Togias' and Plaut's authorships do not constitute endorsement by the US National Institute of Allergy and Infectious Diseases or by any other United States government agency.

References

1. Macdonald, S.M.; Lichtenstein, L.M.; Proud, D.; Plaut, M.; Naclerio, R.M.; MacGlashan, D.W.; Kagey-Sobotka, A. Studies of IgE-dependent histamine releasing factors: Heterogeneity of IgE. *J. Immunol.* **1987**, *139*, 506–512. [PubMed]
2. MacDonald, S.; Rafnar, T.; Langdon, J.; Lichtenstein, L. Molecular identification of an IgE-dependent histamine-releasing factor. *Science* **1995**, *269*, 688–690. [CrossRef] [PubMed]
3. Kawakami, Y.; Kasakura, K.; Kawakami, T. Histamine-Releasing Factor, a New Therapeutic Target in Allergic Diseases. *Cells* **2019**, *8*, 1515. [CrossRef] [PubMed]
4. MacDonald, S.M. History of Histamine-Releasing Factor (HRF)/Translationally Controlled Tumor Protein (TCTP) Including a Potential Therapeutic Target in Asthma and Allergy. *Results Probl. Cell Differ.* **2017**, *64*, 291–308. [CrossRef] [PubMed]

MDPI
St. Alban-Anlage 66
4052 Basel
Switzerland
Tel. +41 61 683 77 34
Fax +41 61 302 89 18
www.mdpi.com

Cells Editorial Office
E-mail: cells@mdpi.com
www.mdpi.com/journal/cells

www.ingramcontent.com/pod-product-compliance
Lightning Source LLC
Chambersburg PA
CBHW062048220926
51850CB00010B/2701
* 9 7 8 3 0 3 6 5 2 2 8 5 2 *